Springer Series in

OPTICAL SCIENCES 110

founded by H.K.V. Lotsch

Springer Series in

OPTICAL SCIENCES

The Springer Series in Optical Sciences, under the leadership of Editor-in-Chief *William T. Rhodes*, Georgia Institute of Technology, USA, provides an expanding selection of research monographs in all major areas of optics: lasers and quantum optics, ultrafast phenomena, optical spectroscopy techniques, optoelectronics, quantum information, information optics, applied laser technology, industrial applications, and other topics of contemporary interest.

With this broad coverage of topics, the series is of use to all research scientists and engineers who need up-to-date reference books.

The editors encourage prospective authors to correspond with them in advance of submitting a manuscript. Submission of manuscripts should be made to the Editor-in-Chief or one of the Editors. See also http://www.springer.de/phys/books/optical_science/

V. Lucarini J.J. Saarinen
K.-E. Peiponen E.M. Vartiainen

Kramers–Kronig Relations in Optical Materials Research

With 37 Figures

Dr. Valerio Lucarini
University of Camerino, Department of Mathematics and Computer Science
Via Madonna delle Carceri 7, 62032 Camerino (MC), Italy
E-mail: valerio.lucarini@unicam.it

Professor Dr. Kai-Erik Peiponen
University of Joensuu, Department of Physics
P.O.Box 111, 80101 Joensuu, Finland
E-mail: kai.peiponen@joensuu.fi

Dr. Jarkko J. Saarinen
University of Toronto, Department of Physics
60 St. George St., Toronto, Ontario M5S 1A7, Canada
E-mail: saarinen@physics.utoronto.ca

Dr. Erik M. Vartiainen
Lappeenranta University of Technology, Department of Electrical Engineering
P.O.Box 20, 53851 Lappeenranta, Finland
E-mail: Erik.Vartiainen@lut.fi

ISSN 0342-4111

ISBN-10 3-540-23673-2 Springer Berlin Heidelberg New York

ISBN-13 978-3-540-23673-2 Springer Berlin Heidelberg New York

Library of Congress Control Number: 2005921306

Springer is a part of Springer Science+Business Media.

springeronline.com

Typesetting and prodcution: PTP-Berlin, Protago-TEX-Production GmbH, Berlin
Cover concept by eStudio Calamar Steinen using a background picture from The Optics Project. Courtesy of John T. Foley, Professor, Department of Physics and Astronomy, Mississippi State University, USA.
Cover production: *design & production* GmbH, Heidelberg

Printed on acid-free paper SPIN: 10977422 57/3141/YU 5 4 3 2 1 0

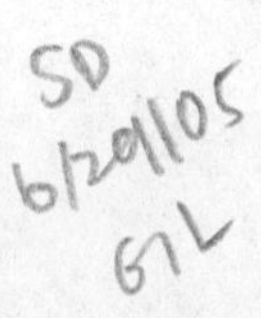

Preface

The Kramers-Kronig relations constitute the mathematical formulation of the fundamental connection between the in-phase to the out-of-phase response of a system to a sinusoidal time-varying external perturbation. Such connection exists in both classical and quantum physical systems and derives directly from the principle of causality.

Apart from being of great importance in high energy physics, statistical physics, and acoustics, at present the Kramers-Kronig relations are basic and widely-accepted tools for the investigation of the linear optical properties of materials, since they allow performing the so-called inversion of optical data, i.e. acquiring knowledge on dispersive phenomena by measurements of absorptive phenomena over the whole energy spectrum or vice versa.

Since the late '80s, a growing body of theoretical results as well as of experimental evidences has shown that the Kramers-Kronig relations can be adopted for efficiently acquiring knowledge on nonlinear optical phenomena. These results suggest that the Kramers-Kronig relations may become in a near future standard techniques in the context of nonlinear spectroscopy.

This book is the first comprehensive treatise devoted to providing a unifying picture of the physical backgrounds, of the rigorous mathematical theory, and of the applications of the Kramers-Kronig relations in both fields of linear and nonlinear optical spectroscopy. Some basic programs written for the MATLAB®[1] environment are also included.

This book is organized as an argumentative discourse, progressing from the linear to the nonlinear phenomena, from the general to the specific systems, and from the theoretical to the experimental results.

This book is intended to all those who are interested and involved in optical materials research such as physicists, chemists, engineers, as well as scientists working in cross-disciplinary fields having contact points with optical materials research.

We hope that this effort will form a baseline for future theoretical and experimental research and technological applications related to wide spectral range nonlinear optical properties.

The authors wish to thank F. Bassani for having encouraged the preparation of this book and S. Scandolo for having provided suggestions on an

[1] MATLAB® is a registered trademark of The MathWorks Inc.

early version. V. L. wishes to thank A. Speranza for continuous support. V. L., J. J. S., and K.-E. P. wish to thank the Academy of Finland for financial support. The financial support of the Finnish Academy of Science and Letters is greatly appreciated.

Joensuu, Camerino, Toronto, and Lappeenranta
November 11, 2004.

Valerio Lucarini
Jarkko J. Saarinen
Kai-Erik Peiponen
Erik M. Vartiainen

Contents

1 **Introduction** 1

2 **Electrodynamic Properties of a General Physical System** 5
- 2.1 The Maxwell Equations 5
- 2.2 The System: Lagrangian and Hamiltonian Descriptions 6
- 2.3 Polarization as a Statistical Property of a System 9

3 **General Properties of the Linear Optical Response** 11
- 3.1 Linear Optical Properties 11
  - 3.1.1 Transmission and Reflection at the Boundary Between Two Media 14
- 3.2 Microscopic Description of Linear Polarization 16
- 3.3 Asymptotic Properties of Linear Susceptibility 17
- 3.4 Local Field and Effective Medium Approximation in Linear Optics 19
  - 3.4.1 Homogeneous Media 19
  - 3.4.2 Two-Phase Media 21

4 **Kramers-Kronig Relations and Sum Rules in Linear Optics** 27
- 4.1 Introductory Remarks 27
- 4.2 The Principle of Causality 27
- 4.3 Titchmarsch's Theorem and Kramers-Kronig Relations 28
  - 4.3.1 Kramers-Kronig Relations for Conductors 29
  - 4.3.2 Kramers-Kronig Relations for the Effective Susceptibility of Nanostructures 30
- 4.4 Superconvergence Theorem and Sum Rules 31
- 4.5 Sum Rules for Conductors 33
  - 4.5.1 Sum Rules for the Linear Effective Susceptibilities of Nanostructures 33
- 4.6 Integral Properties of Optical Constants 34
  - 4.6.1 Integral Properties of the Index of Refraction 35
  - 4.6.2 Kramers-Kronig Relations in Linear Reflectance Spectroscopy 39

4.7 Generalization of Integral Properties for More Effective Data Analysis ... 44
4.7.1 Generalized Kramers-Kronig Relations ... 45
4.7.2 Subtractive K-K Relations ... 47

5 General Properties of the Nonlinear Optical Response ... 49
5.1 Nonlinear Optics: A Brief Introduction ... 49
5.2 Nonlinear Optical Properties ... 51
5.2.1 Pump-and-Probe Processes ... 54
5.3 Microscopic Description of Nonlinear Polarization ... 56
5.4 Local Field and Effective Medium Approximation in Nonlinear Optics ... 58
5.4.1 Homogeneous Media ... 58
5.4.2 Two-Phase Media ... 60
5.4.3 Tailoring of the Optical Properties of Nanostructures ... 63

6 Kramers-Kronig Relations and Sum Rules in Nonlinear Optics ... 71
6.1 Introductory Remarks ... 71
6.2 Kramers-Kronig Relations in Nonlinear Optics: Independent Variables ... 72
6.3 Scandolo's Theorem and Kramers-Kronig Relations in Nonlinear Optics ... 73
6.4 Kramers-Kronig Analysis of the Pump-and-Probe System ... 77
6.4.1 Generalization of Kramers-Kronig Relations and Sum Rules ... 79

7 Kramers-Kronig Relations and Sum Rules for Harmonic-Generation Processes ... 83
7.1 Introductory Remarks ... 83
7.2 Application of Scandolo's Theorem to Harmonic-Generation Susceptibility ... 83
7.3 Asymptotic Behavior of Harmonic-Generation Susceptibility . 84
7.4 General Kramers-Kronig Relations and Sum Rules for Harmonic-Generation Susceptibility ... 87
7.4.1 General Integral Properties of Nonlinear Conductors .. 89
7.5 Subtractive Kramers-Kronig Relations for Harmonic-Generation Susceptibility ... 90

8 Kramers-Kronig Relations and Sum Rules for Data Analysis: Examples ... 93
8.1 Introductory Remarks ... 93
8.2 Applications of Kramers-Kronig Relations for Data Inversion ... 93
8.2.1 Kramers-Kronig Inversion of Harmonic-Generation Susceptibility ... 94

8.2.2 Kramers-Kronig Inversion of the Second Power of Harmonic-Generation Susceptibility 96
8.3 Verification of Sum Rules for Harmonic-Generation Susceptibility 98
8.4 Application of Singly Subtractive Kramers-Kronig Relations 101
8.5 Estimates of the Truncation Error in Kramers-Kronig Relations 104
8.6 Sum Rules and Static Second-Order Nonlinear Susceptibility 106

9 Modified Kramers-Kronig Relations in Nonlinear Optics 109
9.1 Modified Kramers-Kronig Relations for a Meromorphic Nonlinear Quantity 109
9.2 Sum Rules for a Meromorphic Nonlinear Quantity 112

10 The Maximum Entropy Method: Theory and Applications 115
10.1 The Theory of the Maximum Entropy Method 115
10.2 The Maximum Entropy Method in Linear Optical Spectroscopy 117
10.2.1 Phase Retrieval from Linear Reflectance............. 117
10.2.2 Study of Surface Plasmon Resonance 120
10.2.3 Misplacement Phase Error Correction in Terahertz Time-Domain Spectroscopy 126
10.3 The Maximum Entropy Method in Nonlinear Optical Spectroscopy 128

11 Conclusions 133

A MATLAB® Programs for Data Analysis 137
A.1 Program 1: Estimation of the Imaginary Part via Kramers-Kronig Relations 137
A.2 Program 2: Estimation of the Real via Kramers-Kronig Relations 139
A.3 Program 3: Self-Consistent Estimate of the Real and Imaginary Parts of Susceptibility 141
A.4 Program 4: Estimation of the Imaginary Part via Singly Subtractive Kramers-Kronig Relations 142
A.5 Program 5: Estimation of the Real Part via Singly Subtractive Kramers-Kronig Relations 143

References 145

Index 159

1 Introduction

Optical spectroscopy has been a fundamental method of investigation in optical material research for a long time thanks to its nondestructive nature and its robust devices, which are commercially available in many cases. Recording of the wavelength-dependent optical spectrum is in general probably the best and most important tool in the analysis of the different constituents of any medium. Spectrophotometers are widely used in basic research in laboratory conditions as well as in industrial sectors, for routine quality assessment. There is an ever-growing demand to estimate linear and nonlinear optical spectra of various bulk and composite materials more accurately, which is made technically possible by the continuous development of novel optoelectronic devices which are used, for instance, as light sources and detectors. For instance, the scientific as well as industrial interest in the optical properties of novel materials such as nanocomposites comes from their widespread presence and relevant importance both in strategic engineering and life sciences sectors. Another nonconventional and very recent area of interest for the methodologies of optical investigation is in the field of bio-optical medicine, where the interaction of light with different species can be exploited in novel drug development.

In the investigation of linear optical properties, a well-known spectroscopic technique is transmission spectroscopy analysis, based on the utilization of the Beer-Lambert law, where material properties are investigated by the inspection of the height, width, and location of spectral peaks. More sophisticated methods of analysis rely, for instance, on the principal component analysis of measured spectra. Unfortunately, a given experimental analysis cannot usually provide all the optical properties of the sample investigated. A typical example is the inability to obtain a measurement of the refractive index of a medium with a transmission spectroscopy setup.

Fortunately, the requirement of space-time causality in the optical response of any medium provides linear optical functions with general properties [1–3] which can be exploited in order to extract the maximum amount of information from the experimental data on the optical properties of the medium.

The Kramers-Kronig (K-K) relations [4, 5] and the sum rules [6–9] constitute the fundamental theoretical tools of general validity which allow us to widen our knowledge of linear optical phenomena. K-K relations describe a fundamental connection between the real and imaginary parts of linear complex optical functions descriptive of light–matter interaction phenomena, such as susceptibility, the dielectric function, the index of refraction, and reflectivity. The real and imaginary parts are not wholly independent but are connected by a special form of Hilbert transforms, which are termed K-K relations. The sum rules are universal constraints that determine the results of integration over the infinite spectral range of the functions descriptive of relevant optical properties of the medium under investigation. Furthermore, the sum rules have had a key role in the initial development of quantum mechanics. Hence, these integral properties provide constraints for checking the self-consistency of experimental or model-generated data. Furthermore, by applying K-K relations, it is possible to perform the so-called inversion of optical data, i.e., to acquire knowledge on dispersive phenomena by measurements of absorptive phenomena over the whole spectrum (e.g., with transmission spectroscopy) or vice versa. [2,10–15]. In reflection spectroscopy, K-K relations couple the measured reflectance and the phase of the reflectivity, thus allowing retrieval of the phase. K-K relations have been traditionally exploited in linear optical spectroscopy for data inversion and phase retrieval from measured spectra of condensed matter, gases, molecules, and liquids. Furthermore, K-K relations have played a relevant role in different fields, such as high-energy physics, acoustics, statistical physics, and signal processing. The application of K-K relation techniques has greatly increased with the widespread adoption of computers at all levels. Actually, much of the present knowledge of the optical properties of media has been gained by combining measurements and K-K relations. In particular, such studies have had an outstanding impact on the development of optoelectronic devices, for instance, in the development of semiconducting materials for producing detectors of electromagnetic radiation

Linear optics provides a complete description of light–matter interaction only in the limit of weak radiation sources. When we consider more powerful radiation sources, the phenomenology of the interaction is much more complex, since entirely new classes of processes can be observed experimentally. Since the 1960s, the advent of laser technology permitted the observation of nonlinear phenomena in optical frequencies, e.g., harmonic-generation, multiphoton absorption, the Kerr effect, and Raman scattering. These phenomena are produced by the "simultaneous" interaction of matter with several photons. Under quite general hypotheses of the validity of the perturbative approach [16,17], the nth-order nonlinear optical properties of any material can be completely described by corresponding nth-order nonlinear susceptibilities [17–19], which are related to the higher order dynamics of an optical system. Recently, theoretical advances have permitted us to frame in very general terms the possibility of establishing K-K relations for nonlinear optical phenomena [20–26] by determining the class of susceptibility functions that obey K-K relations and sum rules up to any order of nonlinearity.

In spite of the ever-increasing scientific and technological relevance of nonlinear optical phenomena, relatively little attention has been paid for a long time to the experimental investigation of K-K relations and sum rules of corresponding nonlinear susceptibilities [27–30]. Research has usually focused on achieving high resolution in both experimental data and theoretical calculations, although these integral properties are especially relevant for experimental investigations of frequency-dependent nonlinear optical properties. In the context of this class of experiments, K-K relations and sum rules could provide information on whether or not a coherent, common picture of the nonlinear properties of the material under investigation is available [31–34]. The technical problem of measuring the nonlinear excitation spectrum on a relatively wide frequency range, which has been relaxed by the improvements in tunable laser technology, is probably the single most important reason why experimental research in this field has been subdued for a long time.

In this book, we present a detailed analysis of linear and nonlinear optical systems and describe the general integral properties of the susceptibility functions descriptive of optical processes. We deduce rigorously very general K-K dispersion relations and sum rules, and we present the generalization of these integral relations for various kinds of optically linear and nonlinear systems, also including nanostructures. We then describe the applications of K-K relations and sum rules to actual measurements for performing the inversion of optical data and for testing the self-consistency of the spectra of nonlinear optical materials. In addition, we consider the utilization of the maximum entropy method in phase retrieval problems such as those in the new field of terahertz spectroscopy. In the appendix, we also present some simple programs written for the MATLAB® environment that may be used for K-K analysis of data.

We adopt the cgs system of units because it entails great simplification in many formulas. A complete report on how to convert equations and amounts between the cgs and the MKS system is given in Appendix 4 of the classical treatise by Jackson [35].

This book is intended for all those who are interested and involved in optical materials research such as physicists, chemists, engineers and also scientists working in cross-disciplinary fields that have contact points with optical materials research.

2 Electrodynamic Properties of a General Physical System

2.1 The Maxwell Equations

Midway through the nineteenth century, Maxwell combined Coulomb's results on the forces between charged particles and Faraday's investigations on the effects of currents and magnetic fields into four partial differential equations, which reveal the electromagnetic nature of light. Maxwell realized that the solution to the equations of electromagnetism can be expressed as a transverse electromagnetic field [36]. Maxwell's hypothesis was confirmed in 1887 by Hertz who was able to produce and to detect electromagnetic waves.

The Maxwell equations relate the space and time derivatives of electric and magnetic fields to each other throughout a continuous medium. In this section, we present only the main definitions and results which are needed for the scope of this work; for a detailed description of electromagnetic theory see, e.g. [37, 38]. If we adopt cgs units, Maxwell equations can be expressed as follows [37]:

$$\nabla \cdot \boldsymbol{D}(\boldsymbol{r},t) = 4\pi\rho(\boldsymbol{r},t), \tag{2.1}$$

$$\nabla \cdot \boldsymbol{B}(\boldsymbol{r},t) = 0, \tag{2.2}$$

$$\nabla \times \boldsymbol{E}(\boldsymbol{r},t) = -\frac{1}{c}\frac{\partial}{\partial t}\boldsymbol{B}(\boldsymbol{r},t), \tag{2.3}$$

$$\nabla \times \boldsymbol{H}(\boldsymbol{r},t) = \frac{4\pi}{c}\boldsymbol{J}(\boldsymbol{r},t) + \frac{1}{c}\frac{\partial}{\partial t}\boldsymbol{D}(\boldsymbol{r},t), \tag{2.4}$$

where $\boldsymbol{r}$ is the three-dimensional coordinate vector and t indicates time. Here $\boldsymbol{D}(\boldsymbol{r},t)$ denotes the electric displacement, $\rho(\boldsymbol{r},t)$ the charge density, $\boldsymbol{B}(\boldsymbol{r},t)$ the magnetic induction, $\boldsymbol{E}(\boldsymbol{r},t)$ the electric field, $\boldsymbol{H}(\boldsymbol{r},t)$ the magnetic field, and $\boldsymbol{J}(\boldsymbol{r},t)$ the current density. Electric displacement and Magnetic induction are connected to electric and magnetic fields, respectively, by the constitutive equations

$$\boldsymbol{D}(\boldsymbol{r},t) = \boldsymbol{E}(\boldsymbol{r},t) + 4\pi\boldsymbol{P}(\boldsymbol{r},t), \tag{2.5}$$

$$\boldsymbol{B}(\boldsymbol{r},t) = \boldsymbol{H}(\boldsymbol{r},t) + 4\pi\boldsymbol{M}(\boldsymbol{r},t), \tag{2.6}$$

where $\boldsymbol{P}(\boldsymbol{r},t)$ and $\boldsymbol{M}(\boldsymbol{r},t)$ are the polarization and magnetization of the medium, respectively. These quantities describe the response of the medium to an applied electromagnetic field. If we apply the Fourier transform [39] for both space and time to the Maxwell equations, we obtain the following set of equations:

$$\boldsymbol{k} \cdot \boldsymbol{D}(\boldsymbol{k}, \omega) = 4\pi\rho(\boldsymbol{k}, \omega), \tag{2.7}$$

$$\boldsymbol{k} \cdot \boldsymbol{B}(\boldsymbol{k}, \omega) = 0, \tag{2.8}$$

$$\boldsymbol{k} \times \boldsymbol{E}(\boldsymbol{k}, \omega) = \frac{\omega}{c}\boldsymbol{B}(\boldsymbol{k}, \omega), \tag{2.9}$$

$$\boldsymbol{k} \times \boldsymbol{H}(\boldsymbol{k}, \omega) = -\frac{4\mathrm{i}\pi}{c}\boldsymbol{J}(\boldsymbol{k}, \omega) - \frac{\omega}{c}\boldsymbol{D}(\boldsymbol{k}, \omega), \tag{2.10}$$

where $\boldsymbol{k}$ and ω are the wave vector and the angular frequency of the electromagnetic field, respectively.

At optical frequencies, materials are usually nonmagnetic and magnetization can be omitted, so that $\boldsymbol{H}(\boldsymbol{k}, \omega) = \boldsymbol{B}(\boldsymbol{k}, \omega)$. Under this fairly good approximation, the optical response of a medium to an electromagnetic perturbation is completely described by the constitutive relation between polarization and the electric field inducing it.

2.2 The System: Lagrangian and Hamiltonian Descriptions

We consider a volume V containing N electrons with charge $-e$ that interact with a time-dependent electromagnetic field and are subjected to a static scalar potential and to mutual repulsion. The nonrelativistic Lagrangian of the system can then be written as [17, 38]

$$L = \sum_{\alpha=1}^{N} m\frac{\dot{\boldsymbol{r}}^\alpha \cdot \dot{\boldsymbol{r}}^\alpha}{2} - \sum_{\alpha=1}^{N} V(\boldsymbol{r}^\alpha) - \sum_{\alpha\neq\beta=1}^{N} \frac{e^2}{|\boldsymbol{r}^\alpha - \boldsymbol{r}^\beta|} - \sum_{\alpha=1}^{N} \frac{e}{c}\dot{\boldsymbol{r}}^\alpha \cdot \boldsymbol{A}(\boldsymbol{r}^\alpha, t), \tag{2.11}$$

where $\boldsymbol{r}$ is the vector denoting the spatial position, $\boldsymbol{A}$ is the vector potential of the external electromagnetic field, α is the index of the single particle, c is the speed of light in vacuum, and the dot indicates the temporal derivative. We recall that using the Coulomb gauge, we can establish the following relation between the external vector potential $\boldsymbol{A}(\boldsymbol{r}, t)$ and the external electric field $\boldsymbol{E}(\boldsymbol{r}, t)$ [35, 38]:

$$\boldsymbol{E}(\boldsymbol{r}, t) = -\frac{1}{c}\frac{\partial}{\partial t}\boldsymbol{A}(\boldsymbol{r}, t). \tag{2.12}$$

Using the conventional Legendre trasformation [40], we obtain

$$\dot{\boldsymbol{r}}^\alpha = \frac{1}{m}\boldsymbol{p}^\alpha - \frac{e}{mc}\boldsymbol{A}(\boldsymbol{r}^\alpha, t). \tag{2.13}$$

The corresponding nonrelativistic Hamiltonian H governing the dynamics of the system is given by the sum of two terms,

$$H = H_0 + H_I, \tag{2.14}$$

where the first term is the unperturbed, time-independent Hamiltonian

$$H_0 = T + V + H_{\mathrm{ee}} = \sum_{\alpha=1}^{N} \frac{\boldsymbol{p}^\alpha \cdot \boldsymbol{p}^\alpha}{2m} + \sum_{\alpha=1}^{N} V(\boldsymbol{r}^\alpha) + \sum_{\alpha\neq\beta=1}^{N} \frac{e^2}{|\boldsymbol{r}^\alpha - \boldsymbol{r}^\beta|}, \tag{2.15}$$

where T, V, and H_{ee} are, respectively, the total kinetic energy, the total one-particle potential energy, and the electron–electron interaction potential energy of the electrons. The second term is the Hamiltonian of the interaction of electrons with an electromagnetic field:

$$H_I = \sum_{\alpha=1}^{N} \left\{ \frac{e}{2mc} \left[\boldsymbol{p}^\alpha \cdot \boldsymbol{A}(\boldsymbol{r}^\alpha, t) + \boldsymbol{A}(\boldsymbol{r}^\alpha, t) \cdot \boldsymbol{p}^\alpha \right] + \frac{e^2 \boldsymbol{A}(\boldsymbol{r}^\alpha, t) \cdot \boldsymbol{A}(\boldsymbol{r}^\alpha, t)}{2mc^2} \right\}, \tag{2.16}$$

where we have symmetrized the scalar product between the vector potential and the conjugate moments [17], since $[\boldsymbol{A}(\boldsymbol{r}^\alpha, t), \boldsymbol{p}^\alpha] \neq 0$, with the usual definition of the commutator $[a, b] = ab - ba$.

The dynamics of the system does not change if we add a total time derivative [40] to the original Lagrangian (2.11), so that a wholly equivalent Lagrangian is

$$\begin{aligned} \tilde{L} &= L + \frac{\mathrm{d}}{\mathrm{d}t} \left[\sum_{\alpha=1}^{N} \frac{e}{c} \dot{\boldsymbol{r}}^\alpha \cdot \boldsymbol{A}(\boldsymbol{r}^\alpha, t) \right] \\ &= \sum_{\alpha=1}^{N} m \frac{\dot{\boldsymbol{r}}^\alpha \cdot \dot{\boldsymbol{r}}^\alpha}{2} - \sum_{\alpha=1}^{N} V(\boldsymbol{r}^\alpha) - \sum_{\alpha\neq\beta=1}^{N} \frac{e^2}{|\boldsymbol{r}^\alpha - \boldsymbol{r}^\beta|} - \sum_{\alpha=1}^{N} \frac{e}{c} \boldsymbol{r}^\alpha \cdot \frac{\mathrm{d}}{\mathrm{d}t} \boldsymbol{A}(\boldsymbol{r}^\alpha, t). \end{aligned} \tag{2.17}$$

If we apply the Legendre transformation, we obtain a new Hamiltonian function $\tilde{H}$, which is fully equivalent to H:

$$\tilde{H} = H_0 + \tilde{H}_I, \tag{2.18}$$

where H_0 is the same as in (2.15), while we have a new form of the Hamiltonian describing the interaction between an external electromagnetic field and electrons:

$$\tilde{H}_I = -\sum_{\alpha=1}^{N} \frac{e}{c} \boldsymbol{r}^\alpha \cdot \frac{\mathrm{d}}{\mathrm{d}t} \boldsymbol{A}(r^\alpha, t). \tag{2.19}$$

The two interaction Hamiltonians (2.16) and (2.19) are then totally equivalent and can be used interchangeably [41].

If we assume dipolar approximation [14], which we will later describe in Sect. 3.2, we can neglect the spatial dependence of the vector potential $\boldsymbol{A}$, so that it is only a function of time. Hence,

$$[\boldsymbol{A}(t), \boldsymbol{p}^{\alpha}] = 0, \quad \forall \alpha. \tag{2.20}$$

We then find that the interaction Hamiltonian (2.16) can be written as

$$H_I^t = \sum_{\alpha=1}^{N} \left[\frac{e}{mc} \boldsymbol{p}^{\alpha} \cdot \boldsymbol{A}(t) + \frac{e^2 \boldsymbol{A}(t) \cdot \boldsymbol{A}(t)}{2mc^2} \right]. \tag{2.21}$$

Expression (2.21) describes light–matter interaction in the so-called *gauge of velocity*.

We wish to emphasize that in the dipolar approximation we can find a very instructive expression for the interaction Hamiltonian (2.19) that involves the conventional dipole–electric field interaction. Considering that in the dipolar approximation, the constitutive relation (2.12) can be written as

$$\boldsymbol{E}(t) = -\frac{1}{c} \partial_t \boldsymbol{A}(t), \tag{2.22}$$

and noting that when spatial dependence is neglected, the total and partial time derivatives can be identified, so that

$$\frac{\mathrm{d}}{\mathrm{d}t} \boldsymbol{A}(t) = \partial_t \boldsymbol{A}(t), \tag{2.23}$$

we find that the interaction Hamiltonian (2.19) can eventually be expressed as

$$\tilde{H}_I^t = e \sum_{\alpha=1}^{N} \boldsymbol{r}^{\alpha} \cdot \boldsymbol{E}(t). \tag{2.24}$$

Expression (2.24) is often referred to as the interaction Hamiltonian in the *gauge of length.* The notation on the temporal dependency of the interaction Hamiltonian functions (2.21) and (2.24) depends on the fact that we want to emphasize the choice of dipolar approximation, which implies that the external fields are only time-dependent.

We emphasize that, even if we are guaranteed that the physics described by the two gauges is exactly the same, the two gauges have very different performance when adopted in calculations for real systems. In fact, the equivalence is respected only if a truly complete set of eigenstates of the unperturbed Hamiltonian is considered for the intermediate states in the calculations. In actual numerical evaluations of the behavior of real systems, it is impossible to use complete sets. Therefore, the issue of evaluating the relative efficiency of the length and velocity gauge for approximate calculations is of crucial importance. Detailed calculations performed on simple atoms shows that the length gauge tends to outperform the velocity gauge and provides better convergence when relatively few intermediate states are considered [42, 43].

2.3 Polarization as a Statistical Property of a System

All statistical properties of a general quantum system of electrons can be described in terms of the density operator ρ [11] defined as

$$\rho = \sum_{a,b} \rho_{ab} |a\rangle\langle b|, \tag{2.25}$$

where the $|a\rangle$s constitute a complete set of normalized ($\langle a_i|a_j\rangle = \delta_{ij}$) eigenstates of the unperturbed, time-independent Hamiltonian H_0 in the Hilbert space of N indistinguishable fermionic particles and the coefficients ρ_{ab} describe the statistical mixture. The density is normalized to 1 by imposing that

$$\mathrm{Tr}\{\rho\} = \sum_{a} \rho_{aa} = 1. \tag{2.26}$$

We define the expectation value of the operator O as

$$\mathrm{Tr}\{O\rho\} = \sum_{a} O_{ab}\rho_{ba}, \tag{2.27}$$

where we have considered the usual definition $O_{ab} = \langle a|O|b\rangle$. The electric polarization can then be defined as the expectation value of the dipole moment per unit volume [11, 18, 19, 35],

$$\boldsymbol{P}(t) = \frac{1}{V}\mathrm{Tr}\left\{\sum_{a} -e\boldsymbol{r}^{\alpha}\rho(t)\right\}, \tag{2.28}$$

where $\rho(t)$ is the evolution of the density matrix of the system at time t, with the initial condition given by the Boltzmann equilibrium distribution,

$$\rho(0) = \frac{\sum_{a} \exp(-E_a/KT)|a\rangle\langle a|}{\sum_{a} \exp(-E_a/KT)}. \tag{2.29}$$

We assume that the material under examination has no permanent electrical dipole at the thermodynamic equilibrium

$$\frac{1}{V}\mathrm{Tr}\left\{\sum_{\alpha} -e\boldsymbol{r}^{\alpha}\rho(0)\right\} = 0. \tag{2.30}$$

In this study, we ignore the effects of natural radiative decay and noise-induced quantum dephasing at a dynamic level [17]. Hence, we consider a purely semiclassical treatment of the light–matter interaction, so that $\rho(t)$ obeys the Liouville differential equation [18], where the relaxation terms are set to zero:

$$i\hbar\partial_t\rho(t) = [H, \rho(t)] = [H_0, \rho(t)] + \left[H_I^t, \rho(t)\right], \tag{2.31}$$

with boundary condition (2.29) at time $t = 0$. We can solve (2.31) using a perturbative approach, and we can write $\rho(t)$ as a sum consisting of terms of decreasing magnitude:

$$\rho(t) = \rho(0) + \rho^{(1)}(t) + \ldots + \rho^{(n)}(t) + \ldots \tag{2.32}$$

Substituting (2.32) in (2.31) and equating the terms of the same order, we obtain a concatenated system of coupled differential equations:

$$\begin{aligned} i\partial_t \rho^{(1)}(t) &= \left[H_0, \rho^{(1)}(t)\right] + \left[H_I^t, \rho(0)\right], \\ &\ldots \\ i\hbar\partial_t \rho^{(n)}(t) &= \left[H_0, \rho^{(n)}(t)\right] + \left[H_I^t, \rho^{(n-1)}(t)\right]. \\ &\ldots \end{aligned} \tag{2.33}$$

Inasmuch as the initial condition of (2.31) is given by (2.30), $\rho^{(n)}(0) = 0$ for each $n > 0$.

Correspondingly, we can express the total polarization $\boldsymbol{P}(t)$ as the sum of terms of decreasing magnitude:

$$\boldsymbol{P}(t) = \sum_{j=1}^{\infty} \boldsymbol{P}^{(j)}(t) = \sum_{j=1}^{\infty} \frac{1}{V} \mathrm{Tr} \left\{ \sum_a -e\boldsymbol{r}^{a} \rho^{(j)}(t) \right\}, \tag{2.34}$$

where the jth term $\boldsymbol{P}^{(j)}(t)$ describes the jth-order nonlinear optical response of the material to an external electromagnetic field.

Note that the validity of the perturbative approach is related to the assumption that the absolute value of an oscillating electric field affecting electrons is much smaller than a characteristic static electric field due to interactions with the nucleus and with other electrons.

3 General Properties of the Linear Optical Response

3.1 Linear Optical Properties

Linear polarization $\boldsymbol{P}^{(1)}(\boldsymbol{r},t)$ provides an extensive description of light–matter interaction when low radiation intensities are considered. We will consider in a later chapter the effects coming into play when we consider more intense light sources. In general, it is possible to express linear polarization by the following convolution [14]:

$$P_i^{(1)}(\boldsymbol{r},t) = \int\limits_{\mathbb{R}^3}\int\limits_{-\infty}^{\infty} G_{ij}^{(1)}(\boldsymbol{r}-\boldsymbol{r}',t-t')E_j^{\mathrm{loc}}(\boldsymbol{r}',t')\mathrm{d}t'\mathrm{d}\boldsymbol{r}', \tag{3.1}$$

where $\boldsymbol{E}^{\mathrm{loc}}(\boldsymbol{r},t)$ is the local electric field induced by an external perturbation acting in location $\boldsymbol{r}$ at time t and the tensor $G_{ij}^{(1)}(\boldsymbol{r},t)$ is the linear Green function accounting for the dynamic response of the system to the presence of an external electric field.

We emphasize that polarization is a macroscopic quantity, which derives from a suitably defined small-scale spatial average of a corresponding microscopic quantity, polarizability [35, 38, 44, 45]. Polarizability describes the distortion of a dipole field on an atomic scale induced by interaction with an external oscillating field. Geometry suggests that the ratio between polarization and polarizability is simply the number of elementary components of the medium per unit volume. This is not precisely the case, since in the definition of the averaged macroscopic quantity, not only the external applied field but also the local dipole fields induced by single elementary components of the medium yield contributions [44–48]. Therefore, the electric field inducing polarization is not the macroscopic external field of incoming radiation. This implies that $\boldsymbol{E}^{\mathrm{loc}}(\boldsymbol{r},t) \neq \boldsymbol{E}(\boldsymbol{r},t)$, where the latter is the external electric field of incoming radiation. The concept of a local field was originally introduced by Lorentz [49]. In the derivation of the main properties of the linear optical response, we will ignore local field corrections and assume that $\boldsymbol{E}^{\mathrm{loc}}(\boldsymbol{r},t) \sim \boldsymbol{E}(\boldsymbol{r},t)$. In Sect. 3.4, we will deal with local field effects and show how their inclusion affect our results.

We compute the Fourier transform in the time and space domains of both members of the expression (3.1), and considering that the Fourier transform of the convolution of two functions is the conventional product of the Fourier transforms of the two functions [39], we obtain

$$\begin{aligned} P_i^{(1)}(\boldsymbol{k},\omega) &= \Phi\left\{\mathrm{F}\left[P_i^{(1)}(\boldsymbol{r},t)\right]\right\} = \Phi\left\{\mathrm{F}\left[G_{ij}^{(1)}(\boldsymbol{r},t)\right]\right\}\Phi\left\{\mathrm{F}\left[E_j(\boldsymbol{r},t)\right]\right\} \\ &= \chi_{ij}^{(1)}(\boldsymbol{k},\omega)\,E_j(\boldsymbol{k},\omega)\,, \end{aligned} \tag{3.2}$$

where Φ and F indicate the application of the Fourier transform in the 3-D space domain and in the time domain, respectively, and where we define the linear susceptibility of the system as

$$\chi_{ij}^{(1)}(\boldsymbol{k},\omega) = \Phi\left\{\mathrm{F}\left[G_{ij}^{(1)}(\boldsymbol{r},t)\right]\right\}. \tag{3.3}$$

If we apply the Fourier transform in the space and time domains to the first constitutive equation presented in (2.5) and consider only the linear effects, we obtain

$$\boldsymbol{D}^{(1)}(\boldsymbol{k},\omega) = \boldsymbol{E}(\boldsymbol{k},\omega) + 4\pi\boldsymbol{P}^{(1)}(\boldsymbol{k},\omega). \tag{3.4}$$

If we insert into (3.4) the result presented in (3.2) for linear polarization, we derive the following expression for each component of the electric displacement vector:

$$D_i^{(1)}(\boldsymbol{k},\omega) = \left[\delta_{ij} + 4\pi\chi_{ij}^{(1)}(\boldsymbol{k},\omega)\right]E_j(\boldsymbol{k},\omega), \tag{3.5}$$

where δ_{ij} is the Kronecker delta, whose value is 1 if $i=j$ and 0 if $i \neq j$. Thus, the constitutive relation between the electric displacement and the electric field can be expressed as follows:

$$D_i^{(1)}(\boldsymbol{k},\omega) = \varepsilon_{ij}(\boldsymbol{k},\omega)E_j(\boldsymbol{k},\omega), \tag{3.6}$$

where $\varepsilon_{ij}(\boldsymbol{k},\omega)$ is the linear dielectric tensor.

If we consider an isotropic medium or a medium with cubic symmetry, the susceptibility (3.3) and the dielectric (3.6) tensors are diagonal in all coordinate systems [37]. Therefore, they can be expressed in the following way:

$$\chi_{ij}^{(1)}(|\boldsymbol{k}|,\omega) = \delta_{ij}\chi^{(1)}(|\boldsymbol{k}|,\omega), \tag{3.7}$$

$$\varepsilon_{ij}(|\boldsymbol{k}|,\omega) = \delta_{ij}\varepsilon(|\boldsymbol{k}|,\omega), \tag{3.8}$$

where $|\boldsymbol{k}|$ is the length of the vector $\boldsymbol{k}$. If the susceptibility tensor is in the form (3.7), we can treat the linear susceptibility and the dielectric function as scalar quantities:

$$\boldsymbol{P}^{(1)}(\boldsymbol{k},\omega) = \chi^{(1)}(|\boldsymbol{k}|,\omega)\,\boldsymbol{E}(\boldsymbol{k},\omega), \tag{3.9}$$

$$\boldsymbol{D}^{(1)}(\boldsymbol{k},\omega) = \varepsilon(|\boldsymbol{k}|,\omega)\,\boldsymbol{E}(\boldsymbol{k},\omega). \tag{3.10}$$

In such a case and if we assume that no sources are present (i.e. $\rho(\boldsymbol{k},\omega) = J_i(\boldsymbol{k},\omega) = 0, \forall i$), it is possible to derive the electromagnetic wave equations straightforwardly [35, 38], thus obtaining for the electric field,

$$|\boldsymbol{k}|^2 \boldsymbol{E}(\boldsymbol{k},\omega) = \varepsilon(|\boldsymbol{k}|,\omega) \frac{\omega^2}{c^2} \boldsymbol{E}(\boldsymbol{k},\omega) . \tag{3.11}$$

This implies that the following dispersion relation holds:

$$|\boldsymbol{k}|^2 = \varepsilon(|\boldsymbol{k}|,\omega) \frac{\omega^2}{c^2} . \tag{3.12}$$

Hence, each monochromatic component of the radiation wave can be written as

$$\boldsymbol{E}_\omega(\boldsymbol{r},t) = \hat{e} E_\omega \exp\left[\mathrm{i}\frac{\omega}{c} \left(\frac{c|\boldsymbol{k}|}{\omega} \hat{k} \cdot \boldsymbol{r} - ct \right) \right] + \text{c.c.} \tag{3.13}$$

$$\boldsymbol{B}_\omega(\boldsymbol{r},t) = \frac{c}{\omega} \boldsymbol{k} \times \boldsymbol{E}_\omega(\boldsymbol{r},t) , \tag{3.14}$$

where the unit vector $\hat{e}$ gives the polarization of light, $\hat{k}$ is the unit vector in the direction of $\boldsymbol{k}$, and E_ω is a constant amplitude. The electric and magnetic fields are orthogonal to $\boldsymbol{k}$, as implied by the two Maxwell equations (2.7)–(2.8) with no sources. The modulus of the wave vector $\boldsymbol{k}$ can be expressed as a function of ω, thanks to the dispersion relation (3.12), so that we obtain

$$\begin{aligned} \boldsymbol{E}_\omega(\boldsymbol{r},t) &= \hat{e} \boldsymbol{E}_\omega \exp\left\{ \mathrm{i}\frac{\omega}{c} \left[\sqrt{\varepsilon(|\boldsymbol{k}|,\omega)} \hat{k} \cdot \boldsymbol{r} - ct \right] \right\} + \text{c.c.} \\ &= \hat{e} \boldsymbol{E}_\omega \exp\left\{ \mathrm{i}\frac{\omega}{c} \left[N(|\boldsymbol{k}|,\omega) \hat{k} \cdot \boldsymbol{r} - ct \right] \right\} + \text{c.c.} \\ &= \hat{e} \boldsymbol{E}_\omega \exp\left\{ \mathrm{i}\frac{\omega}{c} \left[\eta(|\boldsymbol{k}|,\omega) \hat{k} \cdot \boldsymbol{r} - ct \right] \right\} \\ &\quad \times \exp\left[-\frac{\omega}{c} \kappa(|\boldsymbol{k}|,\omega) \hat{k} \cdot \boldsymbol{r} \right] + \text{c.c.}, \end{aligned} \tag{3.15}$$

where we have introduced the customary notation for the index of refraction [35],

$$\begin{aligned} N(|\boldsymbol{k}|,\omega) &= \sqrt{\varepsilon(|\boldsymbol{k}|,\omega)} \\ &= \mathrm{Re}\left\{ \sqrt{1 + 4\pi\chi^{(1)}(|\boldsymbol{k}|,\omega)} \right\} + \mathrm{i}\,\mathrm{Im}\left\{ \sqrt{1 + 4\pi\chi^{(1)}(|\boldsymbol{k}|,\omega)} \right\} \\ &= \eta(|\boldsymbol{k}|,\omega) + \mathrm{i}\kappa(|\boldsymbol{k}|,\omega). \end{aligned} \tag{3.16}$$

Here Re and Im indicate the real and imaginary parts, respectively. The index of refraction, defined as the square root of a complex function of the variable ω, is in general a complex function, so that we can define its real and imaginary parts. The real part of the index of refraction, $\eta(|\boldsymbol{k}|,\omega)$, is responsible for dispersive optical phenomena, since (3.15) implies that phase velocity of a traveling wave (3.13) is $v(|\boldsymbol{k}|,\omega) = c/\eta(|\boldsymbol{k}|,\omega)$. The imaginary part, $\kappa(|\boldsymbol{k}|,\omega)$, is related to the phenomena of light absorption, since it introduces a nonoscillatory real exponential which describes an extinction process. Using definition (3.16), we can find the following equations relating the real and imaginary parts of the index of refraction and of susceptibility

$$\mathrm{Re}\{\varepsilon(|\boldsymbol{k}|,\omega)\} = [\eta(|\boldsymbol{k}|,\omega)]^2 - [\kappa(|\boldsymbol{k}|,\omega)]^2, \tag{3.17}$$

$$\mathrm{Im}\{\varepsilon(|\boldsymbol{k}|,\omega)\} = 2\eta(|\boldsymbol{k}|,\omega)\kappa(|\boldsymbol{k}|,\omega). \tag{3.18}$$

3.1.1 Transmission and Reflection at the Boundary Between Two Media

Suppose that a monochromatic electromagnetic plane wave with frequency ω of the form (3.15) arrives at the boundary of two media with refractive indices $N_1(|\boldsymbol{k_1}|,\omega)$ and $N_2(|\boldsymbol{k_2}|,\omega)$ at $x = 0$, as presented in Fig. 3.1. The structure is assumed to be invariant in the y-direction, and the components of the electromagnetic wave can be expressed in the form

$$\boldsymbol{E} = \boldsymbol{E}_{\mathrm{i/r/t}}(x,z,t) = \boldsymbol{E}_{\mathrm{i/r/t}} \exp\left[\mathrm{i}\left(k_x^{\mathrm{i/r/t}} \cdot x + k_z^{\mathrm{i/r/t}} \cdot z - \omega \cdot t\right)\right], \tag{3.19}$$

$$\boldsymbol{H} = \boldsymbol{H}_{\mathrm{i/r/t}}(x,z,t) = \boldsymbol{H}_{\mathrm{i/r/t}} \exp\left[\mathrm{i}\left(k_x^{\mathrm{i/r/t}} \cdot x + k_z^{\mathrm{i/r/t}} \cdot z - \omega \cdot t\right)\right], \tag{3.20}$$

where the indices i, r, and t refer to the incoming, reflected, and transmitted fields, respectively. Starting with the Maxwell equations, two different cases can be considered. In the case of TE-polarized light, the electric field oscillates perpendicularly to the plane of incident and reflected beams, i.e., in the y-direction. In the case of TM-polarized light, the electric field oscillates in the plane (x,z) of the beams. Obviously, at normal incidence, the TE/TM decomposition is lost. The angle of the reflected beam equals the angle of incidence, and the angle of the refracted beam can be derived from the following equation:

$$N_1(|\boldsymbol{k_1}|,\omega)\sin\varphi_i = N_2(|\boldsymbol{k_2}|,\omega)\sin\varphi_t, \tag{3.21}$$

which can be derived from the boundary conditions [35] and constitutes the generalization of the Snell law to the complex case.

In the case of TE-polarized light, the boundary conditions imply that the y-component of the electric field and the z-component of the magnetic field are continuous. These conditions can be expressed as

$$|\boldsymbol{E}_{\mathbf{i}}| + |\boldsymbol{E}_{\mathbf{r}}| = |\boldsymbol{E}_{\mathbf{t}}|, \tag{3.22}$$

$$|\boldsymbol{E}_{\mathbf{i}}|\, N_1(|\boldsymbol{k_1}|,\omega)\cos\varphi_1 + |\boldsymbol{E}_{\mathbf{r}}|\, N_1(|\boldsymbol{k_1}|,\omega)\cos\varphi_1 = |\boldsymbol{E}_{\mathbf{t}}|\, N_2(|\boldsymbol{k_2}|,\omega)\cos\varphi_2. \tag{3.23}$$

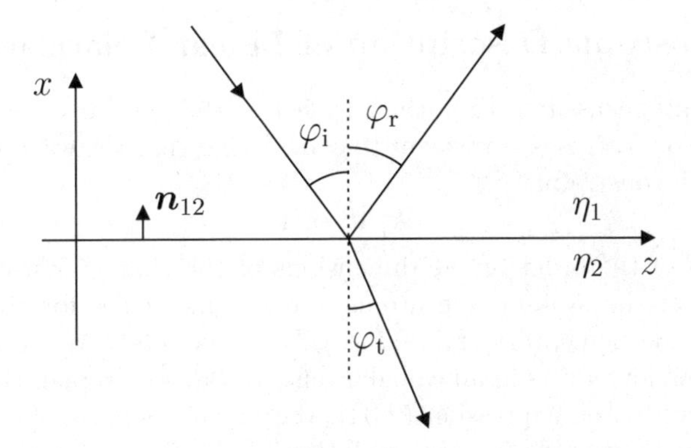

Fig. 3.1. Geometry for the derivation of Fresnel's equations

Using the fact that $\varphi_\mathrm{i} = \varphi_\mathrm{r}$, it is possible to obtain Fresnel's amplitude reflection coefficient for TE-polarized light as

$$r_\mathrm{TE}(\omega) = \left(\frac{|\boldsymbol{E}_\mathbf{r}|}{|\boldsymbol{E}_\mathbf{i}|}\right)_\mathrm{TE} = \frac{N_1(|\boldsymbol{k_1}|,\omega)\cos\varphi_\mathrm{i} - N_2\left(|\boldsymbol{k_2}|,\omega\right)\cos\varphi_\mathrm{t}}{N_1(|\boldsymbol{k_1}|,\omega)\cos\varphi_\mathrm{i} + N_2\left(|\boldsymbol{k_2}|,\omega\right)\cos\varphi_\mathrm{t}}, \tag{3.24}$$

and the amplitude transmission coefficient as

$$t_\mathrm{TE}(\omega) = \left(\frac{|\boldsymbol{E}_\mathbf{t}|}{|\boldsymbol{E}_\mathbf{i}|}\right)_\mathrm{TE} = \frac{2N_1(|\boldsymbol{k_1}|,\omega)\cos\varphi_\mathrm{i}}{N_1(|\boldsymbol{k_1}|,\omega)\cos\varphi_\mathrm{i} + N_2\left(|\boldsymbol{k_2}|,\omega\right)\cos\varphi_\mathrm{t}}. \tag{3.25}$$

For TM-polarized light, the z-component of the electric field and the y-component of the magnetic field are continuous.These conditions can be expressed as

$$|\boldsymbol{E}_\mathbf{i}|\cos\varphi_\mathrm{i} - |\boldsymbol{E}_\mathbf{r}|\cos\varphi_\mathrm{r} = |\boldsymbol{E}_\mathbf{t}|\cos\varphi_\mathrm{t}, \tag{3.26}$$

$$|\boldsymbol{E_1}|\,N_1(|\boldsymbol{k_1}|,\omega) + |\boldsymbol{E}_\mathbf{r}|\,N_1(|\boldsymbol{k_1}|,\omega) = |\boldsymbol{E}_\mathbf{t}|\,N_2(|\boldsymbol{k_1}|,\omega). \tag{3.27}$$

The amplitude reflection and transmission coefficients for TM-polarized light result in the following:

$$r_\mathrm{TM}(\omega) = \left(\frac{|\boldsymbol{E}_\mathbf{r}|}{|\boldsymbol{E}_\mathbf{i}|}\right)_\mathrm{TM} = \frac{N_2\left(|\boldsymbol{k_2}|,\omega\right)\cos\varphi_\mathrm{i} - N_1(|\boldsymbol{k_1}|,\omega)\cos\varphi_\mathrm{t}}{N_2\left(|\boldsymbol{k_2}|,\omega\right)\cos\varphi_\mathrm{i} + N_1(|\boldsymbol{k_1}|,\omega)\cos\varphi_\mathrm{t}}, \tag{3.28}$$

$$t_\mathrm{TM}(\omega) = \left(\frac{|\boldsymbol{E}_\mathbf{t}|}{|\boldsymbol{E}_\mathbf{i}|}\right)_\mathrm{TM} = \frac{2N_1(|\boldsymbol{k_1}|,\omega)\cos\varphi_\mathrm{i}}{N_1(|\boldsymbol{k_1}|,\omega)\cos\varphi_\mathrm{i} + N_2\left(|\boldsymbol{k_2}|,\omega\right)\cos\varphi_\mathrm{t}}. \tag{3.29}$$

These results are traditionally known as Fresnel's equations and are used in calculations of the reflectance and transmittance through the boundaries. Furthermore, they are used to take into account the possible phase shifts occurring at the boundary.

3.2 Microscopic Description of Linear Polarization

In this work, we assume that the physics of the system does not depend appreciably on $\boldsymbol{k}$; this is equivalent to considering that the wave vectors obey the following constraint:

$$|\boldsymbol{k}|\,|\boldsymbol{d}| \ll 1, \tag{3.30}$$

where $|\boldsymbol{d}|$ is of the order of the dimensions of the characteristic elementary constituent of the system, the atoms and the molecules for the gases and liquids, and the elementary crystalline cells for the solids. We then ignore the spatial dispersion effects in all calculations, so that we drop all the $\boldsymbol{r}$ in (3.1).

According to the expression (2.34), the linear response of a system can be described by considering the evolution of the first-order term $\rho^{(1)}(t)$ in the system of differential equations (2.33). This is a first-order differential equation that can be solved with the method of the variation of arbitrary constants. We follow [18] and use the length gauge formulation (2.24) for the interaction Hamiltonian. We then derive the following expression for linear polarization:

$$\begin{aligned} P_i^{(1)}(t) &= \int_{-\infty}^{\infty} G_{ij}^{(1)}(\tau)E_j(t-\tau)\mathrm{d}\tau \\ &= -\frac{e^2}{\mathrm{i}\hbar V}\int_{-\infty}^{\infty} E_j(t-\tau)\theta(\tau)\mathrm{Tr}\left\{\left[\sum_{\alpha=1}^{N} r_j^{\alpha}(-\tau)\sum_{\alpha=1}^{N} r_i^{\alpha}\right]\rho(0)\right\}\mathrm{d}\tau, \end{aligned} \tag{3.31}$$

where we have introduced the usual Heaviside function $\theta(t)$ and obtained the following expression for the linear Green function of the system:

$$G_{ij}^{(1)}(t) = -\frac{e^2}{\mathrm{i}\hbar V}\theta(t)\mathrm{Tr}\left\{\left[\sum_{\alpha=1}^{N} r_j^{\alpha}(-t)\sum_{\alpha=1}^{N} r_i^{\alpha}\right]\rho(0)\right\}, \tag{3.32}$$

which is nil if $t < 0$. We have used the interaction representation for the evolution of the position operators

$$r_j^{\alpha}(-t) = \exp\left[\frac{\mathrm{i}H_0(-t)}{\hbar}\right] r_j^{\alpha} \exp\left[-\frac{\mathrm{i}H_0(-t)}{\hbar}\right]. \tag{3.33}$$

Polarization is the convolution of two functions depending only on time variables, and therefore its Fourier transform is the ordinary product of the temporal Fourier transforms of the two functions

$$P_i^{(1)}(\omega) = \mathrm{F}\left[P_i^{(1)}(t)\right] = \chi_{ij}^{(1)}(\omega)E_j(\omega). \tag{3.34}$$

Expression (3.34) corresponds to the limit $|k| \to 0$ of expression (3.2). We note that the adoption of the dipolar approximation implies that all linear optical functions deriving from linear susceptibility or, equivalently, from the linear dielectric function, are taken as dependent only on the variable ω.

3.3 Asymptotic Properties of Linear Susceptibility

We emphasize that the semiclassical approach we have adopted in this work does not permit a detailed representation of absorption phenomena because evolution in our treatment is driven by purely Hermitian operators. As is well known [14, 18, 19, 50], this approximation implies that absorption peaks are essentially Dirac δ-functions centered on transition frequencies. In a more precise picture, the δ-functions are substituted by Lorentzian functions when a finite lifetime is introduced. The asymptotic behavior of total susceptibility turns out to be related only to the real part since the imaginary part decreases much faster with frequency.

From the standard Fourier transform theory, we see that

$$\mathrm{F}\left[\frac{(-t)^k}{k!}\theta(t)\right] = \int_{-\infty}^{\infty} \frac{(-t)^k}{k!}\theta(t)e^{\mathrm{i}\omega t}\mathrm{d}t = \frac{\mathrm{i}^k}{k!}\frac{\mathrm{d}^k}{\mathrm{d}\omega^k}\left[\mathrm{iP}\frac{1}{\omega} + \pi\delta(\omega)\right], \quad (3.35)$$

where P indicates that the principal part is considered. If we consider asymptotic behavior for large values of ω, we can drop $\delta(\omega)$ and its derivatives as well as the principal part, because they are relevant only for $\omega = 0$, thus obtaining

$$\mathrm{F}\left[\frac{(-t)^k}{k!}\theta(t)\right] \approx -(-\mathrm{i})^{k+1}\frac{1}{\omega^{k+1}}. \quad (3.36)$$

This implies that in order to deduce information on the asymptotic behavior of susceptibility, we must analyze the short-term behavior of the response function. Considering that by definition the relaxation processes are not important on the shortest timescales of the system, we have a relevant conceptual argument that supports the assumption that a purely Hermitian Hamiltonian operator does not affect the leading asymptotic behavior of linear susceptibility.

From the Heisenberg equations, we can deduce that for short times, the following Taylor expansion holds:

$$r_j^\alpha(-t) = \sum_{k=0}^{\infty} a_{k,j}^\alpha \frac{(-t)^k}{k!}, \quad (3.37)$$

where

$$a_{k,j}^\alpha = -\left(\frac{1}{\mathrm{i}\hbar}\right)\left[H_0, a_{k-1,j}^\alpha\right], \quad a_{0,j}^\alpha = r_j^\alpha \; . \quad (3.38)$$

Substituting expression (3.37) in the linear Green function (3.32) and adding on the index α, we obtain the following infinite Taylor expansion:

$$G_{ij}^{(1)}(t) = \sum_{k=0}^{\infty} A_{ij}^k \theta(t)\frac{(-t)^k}{k!}, \quad (3.39)$$

with the following definition for the tensorial coefficients:

$$A_{ij}^{k} = -\frac{e^2}{\mathrm{i}\hbar V}\theta(t)\mathrm{Tr}\left\{\left[\sum_{\alpha=1}^{N} a_{k,j}^{\alpha}, \sum_{\alpha=1}^{N} r_i^{\alpha}\right], \rho(0)\right\}. \tag{3.40}$$

Applying the Fourier transform to the Taylor expansion of the Green function (3.39) and considering the result (3.36), we obtain the result that for large values of ω, the linear susceptibility can be written as

$$\begin{aligned}\chi_{ij}^{(1)}(\omega) &\approx \frac{e^2}{\mathrm{i}\hbar V}\sum_{k=0}^{\infty}\mathrm{Tr}\left\{\left[\sum_{\alpha=1}^{N} a_{k,j}^{\alpha}, \sum_{\alpha=1}^{N} r_i^{\alpha}\right]\rho_0\right\}(-\mathrm{i})^{k+1}\frac{1}{\omega^{k+1}} \\ &= \sum_{k=0}^{\infty} A_{ij}^{k}(-\mathrm{i})^{k+1}\frac{1}{\omega^{k+1}}.\end{aligned} \tag{3.41}$$

Therefore, the leading term in the asymptotic expansion of the linear susceptibility can be found by determining which is the lowest value of k such that the tensorial coefficient (3.40) does not vanish.

When considering $k = 0$ in expression (3.40), we see that

$$A_{ij}^{0} = -\frac{e^2}{\mathrm{i}\hbar V}\mathrm{Tr}\left\{\left[\sum_{\alpha=1}^{N} r_j^{\alpha} \sum_{\alpha=1}^{N} r_i^{\alpha}\right], \rho_0\right\} = 0, \tag{3.42}$$

since it involves the commutator of two purely spatial variables. The value of the $k = 1$ coefficient in expression (3.40) is

$$\begin{aligned}A_{ij}^{1} &= \frac{e^2}{\mathrm{i}\hbar V}\mathrm{Tr}\left\{\left[\frac{1}{\mathrm{i}\hbar}\left[H_0, \sum_{\alpha=1}^{N} r_j^{\alpha}\right]\sum_{\alpha=1}^{N} r_i^{\alpha}\right], \rho_0\right\} \\ &= \frac{e^2}{\mathrm{i}\hbar V}\mathrm{Tr}\left\{\left[-\sum_{\alpha=1}^{N}\frac{p_j^{\alpha}}{m}, \sum_{\alpha=1}^{N} r_i^{\alpha}\right]\rho_0\right\} \\ &= \frac{e^2 N}{Vm}\mathrm{Tr}\left\{\delta_{ij}\rho_0\right\} = \frac{e^2 N}{mV}\delta_{ij} = \frac{\omega_{\mathrm{p}}^2}{4\pi}\delta_{ij},\end{aligned} \tag{3.43}$$

where we have considered that the trace of the density matrix is 1 and we have introduced the plasma frequency with the usual notation ω_{p}. Therefore, we find that, asymptotically, the susceptibility tensor is diagonal and decreases as ω^{-2}:

$$\chi_{ij}^{(1)}(\omega) \approx -\frac{\omega_{\mathrm{p}}^2}{4\pi}\delta_{ij}\frac{1}{\omega^2} + o(\omega^{-2}), \tag{3.44}$$

where $o(\omega^{-2})$ indicates all terms having an asymptotic decrease strictly faster than ω^{-2}.

We can observe that the asymptotic term in (3.44) is isotropic and does not contain physical parameters other than the electron charge, mass, and density, thus excluding the quantum parameter $\hbar$. This implies that, if the material interacts with high-frequency radiation, it behaves universally like a classical free electron gas with the same density.

3.4 Local Field and Effective Medium Approximation in Linear Optics

The response of a medium to an external electric field cannot be described exactly by means of macroscopic electric fields. The external field drives the bound charges of the medium apart and induces a collection of dipole moments [51]. In an optically dense medium, interaction of the induced dipoles is taken into account by a local field factor, which relates the macroscopic fields to the local ones.

Conventionally, the local field is considered by starting from the macroscopic properties of the medium (see, for instance, [44]). Unfortunately, such an approach lacks generality and does not provide any insight into the physical principles responsible for the dielectric properties of the medium. Alternatively, it is possible to arrive at macroscopic properties by averaging the microscopic response over the volume investigated [46]. Recently, a rigorous derivation of the local field corrections was provided with the aid of particle correlations [49]. Furthermore, this rigorous proof is valid for effective dielectric functions of mixtures as well.

For our purposes, the macroscopic approach is suitable. In order to avoid a cumbersome presentation, we have chosen to develop a scalar theory, which corresponds to treating isotropic matter. Nevertheless, it can be immediately seen that the very same conclusions apply to a full tensorial theory.

3.4.1 Homogeneous Media

For materials with linear optical responses, the local field determines the microscopic polarization $\boldsymbol{p}$, expressed as the product of the local electric field times the polarizability $\alpha(\omega)$ of a single microscopic constituent:

$$\boldsymbol{p}(\omega) = \alpha(\omega)\boldsymbol{E}^{\mathrm{loc}}(\omega). \tag{3.45}$$

Macroscopic polarization of the medium is derived by averaging (3.45) over the volume V investigated, as follows [46]:

$$\boldsymbol{P}(\omega) = \frac{1}{V}\int_V \boldsymbol{p}(\omega)\mathrm{d}V = \aleph\alpha(\omega)\boldsymbol{E}_{\mathrm{loc}}(\omega), \tag{3.46}$$

where $\mathrm{d}V$ is the unit volume for integration and $\aleph$ is the density of microscopic constituents. We emphasize that if we consider a homogeneous medium,

$$n_e\aleph = \frac{N}{V}, \tag{3.47}$$

where N/V is the number of electrons per unit volume and n_{e} is the number of electrons per elementary constituent of the medium. On the other hand, polarization can be expressed in terms of the external electromagnetic field as follows:

$$\boldsymbol{P}(\omega) = \chi_{\text{eff}}^{(1)}(\omega)\boldsymbol{E}(\omega). \tag{3.48}$$

Hence, in order to express the effective susceptibility in terms of the microscopic polarizability, we have to express the local electric field in terms of the external electric field. The local electric field can be expressed as follows [44]:

$$\boldsymbol{E}^{\text{loc}}(\omega) = \boldsymbol{E}(\omega) + \frac{4\pi}{3}\boldsymbol{P}(\omega). \tag{3.49}$$

By inserting the definition of macroscopic polarization (3.46) into (3.49), we obtain

$$\begin{aligned} \boldsymbol{E}^{\text{loc}}(\omega) &= \boldsymbol{E}(\omega) + \frac{4\pi}{3}\aleph\alpha(\omega)\boldsymbol{E}^{\text{loc}}(\omega) \\ &= \frac{1}{3}\{3\boldsymbol{E}(\omega) + [\varepsilon(\omega) - 1]\boldsymbol{E}(\omega)\} = \frac{\varepsilon(\omega) + 2}{3}\boldsymbol{E}(\omega), \end{aligned} \tag{3.50}$$

The combination of (3.4)–(3.6) and (3.50) and the use of the definition of the local field yields the result

$$\frac{4\pi}{3}\aleph\alpha(\omega) = \frac{\varepsilon(\omega) - 1}{\varepsilon(\omega) + 2}, \tag{3.51}$$

which is known as the classical Clausius-Mossotti equation [35, 38]. The local field factor corrects the values $\varepsilon(\omega) - 1$ that are calculated without the presence of a local field, e.g., in dilute gases [17].

We can redefine the constitutive relation between induced macroscopic polarization $P^{(1)}(\omega)$ and applied external field $E(\omega)$ presented in (3.34) as the following:

$$\boldsymbol{P}^{(1)}(\omega) = \chi_{\text{eff}}^{(1)}(\omega)\boldsymbol{E}(\omega), \tag{3.52}$$

where the linear macroscopic susceptibility $\chi_{\text{eff}}^{(1)}(\omega)$ is related to the linear microscopic polarizability $\alpha(\omega)$ by the following equation descriptive of local field effects, which is equivalent to expression (3.51):

$$\chi_{\text{eff}}^{(1)}(\omega) = \frac{\aleph\alpha(\omega)}{1 - \frac{4\pi}{3}\aleph\alpha(\omega)}. \tag{3.53}$$

If the density $\aleph$ is small, $\chi_{\text{eff}}^{(1)}(\omega) \sim \aleph\alpha(\omega)$. The linear susceptibility $\chi^{(1)}(\omega)$ presented in (3.41) is linear with density, so that it precisely obeys this approximation:

$$\chi^{(1)}(\omega) = \aleph\alpha(\omega). \tag{3.54}$$

Therefore, we can establish the following functional dependence between the effective macroscopic susceptibility and the susceptibility we computed in (3.41)

$$\chi_{\text{eff}}^{(1)}(\omega) = \frac{\chi^{(1)}(\omega)}{1 - \frac{4\pi}{3}\chi^{(1)}(\omega)}. \tag{3.55}$$

Inserting the asymptotic behavior of susceptibility (3.44) into (3.55), we deduce that $\chi^{(1)}_{\rm eff}(\omega)$ and $\chi^{(1)}(\omega)$ are asymptotically equivalent:

$$\chi^{(1)}_{\rm eff}(\omega) \approx \chi^{(1)}(\omega) \approx -\frac{\omega_p^2}{4\pi}\frac{1}{\omega^2} + o(\omega^{-2}). \tag{3.56}$$

3.4.2 Two-Phase Media

The classical Clausius-Mossotti equation (3.51) can be extended to a first-order effective-medium approximation by considering a mixture of two constituents. For a two-component system, where the constituents have different polarizabilities $\alpha_{\rm a}(\omega)$ and $\alpha_{\rm b}(\omega)$,

$$\frac{\varepsilon_{\rm eff}(\omega) - 1}{\varepsilon_{\rm eff}(\omega) + 2} = \frac{4\pi}{3}\left[\aleph_{\rm a}\alpha_{\rm a}(\omega) + \aleph_{\rm b}\alpha_{\rm b}(\omega)\right], \tag{3.57}$$

where $\aleph_i$ is the number of constituent dipoles per unit volume with i = a or b. Both constituents can be expressed with the aid of (3.51):

$$\frac{\varepsilon_{\rm eff}(\omega) - 1}{\varepsilon_{\rm eff}(\omega) + 2} = f_{\rm a}\frac{\varepsilon_{\rm a}(\omega) - 1}{\varepsilon_{\rm a}(\omega) + 2} + f_{\rm b}\frac{\varepsilon_{\rm b}(\omega) - 1}{\varepsilon_{\rm b}(\omega) + 2}, \tag{3.58}$$

where $f_i = \aleph_i/(\aleph_{\rm a} + \aleph_{\rm b})$ denotes the volume fraction of constituent i. Equation (3.58) is the effective medium approximation, which can be used in the derivation of the optical properties of nanostructures.

The history of research on the optical properties of nanostructures dates back to the studies of Maxwell Garnett [52, 53], who explained the colors induced by minute metal spheres by the effective medium theory. For example, gold nanospheres embedded in glass display a ruby-red color due to an anomalous absorption band [54]. Classical electromagnetism related to randomly oriented nanostructures is concerned with the effective medium approximation originally provided by Bruggeman [55] (see also Landauer [56]). The size of the constituents is assumed to be much smaller than the wavelength of the incident light but large enough that they possess macroscopic dielectric constants.

The Maxwell Garnett Nanosphere System

Probably the simplest case of an effective medium is the two-phase Maxwell Garnett (MG) nanosphere system, presented in Fig. 3.2a. Solid nanospheres are embedded in a host material, and they can be insulators, metals, or semiconductors and are assumed to have a size much smaller than the wavelength of the incident light. Here we consider the case of dielectric nanospheres. In the case of a relatively low volume fraction $f_{\rm i}$ of the nanospheres embedded in an optically linear system, it holds for the linear effective dielectric function $\varepsilon_{\rm eff}$ that

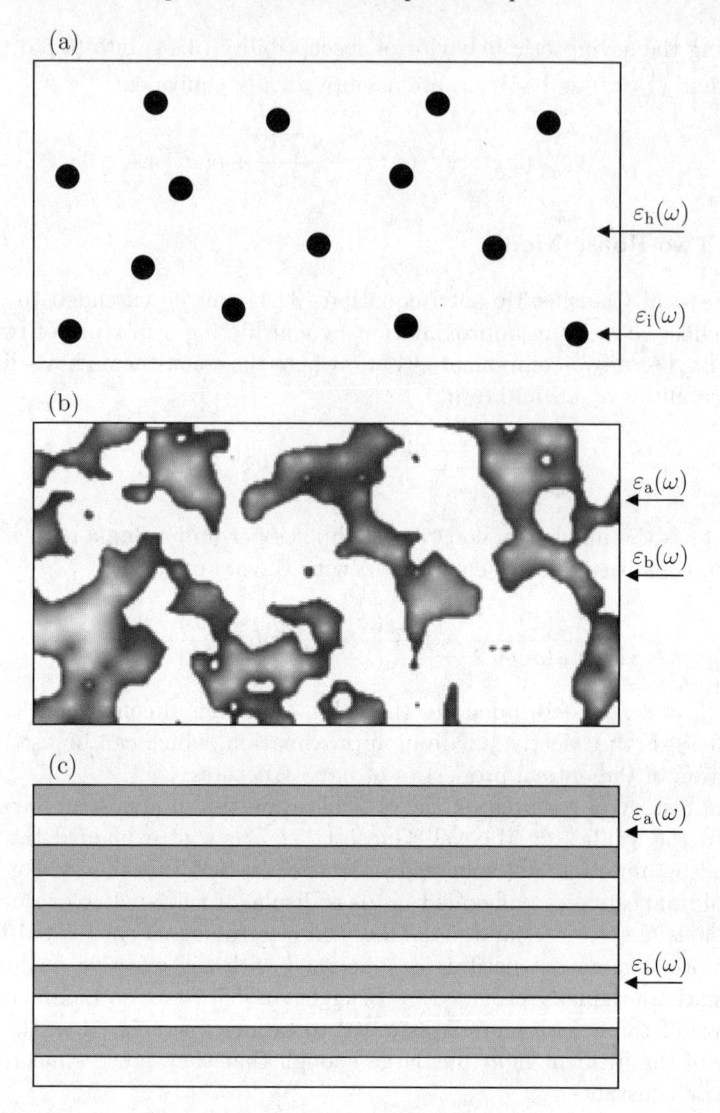

Fig. 3.2. Schematic diagrams of two-phased (**a**) Maxwell Garnett, (**b**) Bruggeman, and (**c**) layered nanostructures

$$\frac{\varepsilon_{\mathrm{eff}}(\omega)-\varepsilon_{\mathrm{h}}(\omega)}{\varepsilon_{\mathrm{eff}}(\omega)+2\varepsilon_{\mathrm{h}}(\omega)}=f_{\mathrm{i}}\frac{\varepsilon_{\mathrm{i}}(\omega)-\varepsilon_{\mathrm{h}}(\omega)}{\varepsilon_{\mathrm{i}}(\omega)+2\varepsilon_{\mathrm{h}}(\omega)}, \tag{3.59}$$

where $\varepsilon_{\mathrm{i}}(\omega)$ is the complex linear dielectric function of the nanospheres and $\varepsilon_{\mathrm{h}}(\omega)$ the corresponding quantity of the host material. Equation (3.59) treats the constituent materials asymmetrically, i.e. the nanospheres are embedded in the host material. The MG approximation of an effective medium [52, 53] can be reformulated as follows:

$$\varepsilon_{\mathrm{eff}}(\omega) = \varepsilon_{\mathrm{h}}(\omega)\frac{1 + 2v(\omega)f_{\mathrm{i}}}{1 - v(\omega)f_{\mathrm{i}}}, \tag{3.60}$$

where $v(\omega)$ is given by

$$v(\omega) = \frac{\varepsilon_{\mathrm{i}}(\omega) - \varepsilon_{\mathrm{h}}(\omega)}{\varepsilon_{\mathrm{i}}(\omega) + 2\varepsilon_{\mathrm{h}}(\omega)}. \tag{3.61}$$

Two major limitations of the MG approximation are that the interaction between nanospheres and the scattering of light are neglected. Furthermore, the MG model does not take into account either the size or the the distance between nanospheres, thus neglecting agglomeration and polydispersion effects. In the literature, some extended versions of the MG approximation, which take into account the size of nanospheres, have been discussed [57,58]. However, in the frame of the conventional MG approximation, the volume fraction of nanospheres must be relatively low ($f_{\mathrm{i}} \ll 1$). Nevertheless, we note that Boyd et al. [59] used a value $f_{\mathrm{i}} = 0.5$ as an upper limit in their theoretical study. In experimental measurements [60], the upper limit of the volume fraction, which gives good agreement between the measured and the predicted values, is rather small ($f_{\mathrm{i}} \approx 0.1$). The increase in the volume fraction shifts the resonance angular frequency of the nanosphere system toward lower energies. An example of this effect in recent experimental studies on nanostructures is the work of Dalacu and Martinu [61], who investigated the spectral shift of the surface plasmon resonance of the nanocomposite films in the frame of the linear optics of the MG model [62].

Unfortunately, direct measurement of the complex effective dielectric function of an MG nanosphere system is not possible in the optical spectral region. However, the reflectance of a liquid-phase MG system can be obtained in order to calculate the complex effective dielectric function of the liquid.

The Bruggeman Effective Medium Theory

In the framework of the MG approximation, the volume fraction of nanospheres is restricted to relatively low values. Bruggeman [55] solved this problem with the assumption that the inclusions are embedded in the effective medium itself and the dielectric function of the host material is replaced by the effective dielectric function in (3.59). Such a procedure yields a symmetric description for the effective dielectric function [55]:

$$f_{\mathrm{a}}\frac{\varepsilon_{\mathrm{a}}(\omega) - \varepsilon_{\mathrm{eff}}(\omega)}{\varepsilon_{\mathrm{a}}(\omega) + 2\varepsilon_{\mathrm{eff}}(\omega)} + f_{\mathrm{b}}\frac{\varepsilon_{\mathrm{b}}(\omega) - \varepsilon_{\mathrm{eff}}(\omega)}{\varepsilon_{\mathrm{b}}(\omega) + 2\varepsilon_{\mathrm{eff}}(\omega)} = 0, \tag{3.62}$$

where a and b denote the two components having different complex dielectric functions ($\varepsilon_{\mathrm{a}}, \varepsilon_{\mathrm{b}}$) and volume fractions ($f_{\mathrm{a}}, f_{\mathrm{b}}$), respectively. The effective dielectric function is invariant if the constituents are replaced by each other, as opposed to the MG case. In the Bruggeman model, presented in Fig. 3.2b, host and guest materials are indistinguishable and the system is completely random. The Bruggeman theory has been generalized in order to account for different shapes of inclusion by Zeng et al. [47], who proposed the following formula:

$$f_{\mathrm{a}} \frac{\varepsilon_{\mathrm{a}}(\omega) - \varepsilon_{\mathrm{eff}}(\omega)}{\varepsilon_{\mathrm{eff}}(\omega) + g[\varepsilon_{\mathrm{a}}(\omega) - \varepsilon_{\mathrm{eff}}(\omega)]} + f_{\mathrm{b}} \frac{\varepsilon_{\mathrm{b}}(\omega) - \varepsilon_{\mathrm{eff}}(\omega)}{\varepsilon_{\mathrm{eff}}(\omega) + g[\varepsilon_{\mathrm{b}}(\omega) - \varepsilon_{\mathrm{eff}}(\omega)]} = 0. \quad (3.63)$$

Here, g is a geometric factor, which depends on the shape of the inclusions. For spherical inclusions, $g = 1/3$ and for two-dimensional circular inclusions, $g = 1/2$. In the case of $g = 1/3$, (3.63) reduces to the original form given by Bruggeman [55]. The quadratic (3.63) has two explicit solutions for the complex effective dielectric function

$$\varepsilon_{\mathrm{eff}}(\omega) = \frac{-c(\omega) \pm \sqrt{c(\omega)^2 + 4g(1-g)\varepsilon_{\mathrm{a}}(\omega)\varepsilon_{\mathrm{b}}(\omega)}}{4(1-g)}, \quad (3.64)$$

where

$$c(\omega) = (g - f_{\mathrm{a}})\varepsilon_{\mathrm{a}}(\omega) + (g - f_{\mathrm{b}})\varepsilon_{\mathrm{b}}(\omega). \quad (3.65)$$

In (3.64), only the positive branch is physically reasonable. This can be justified by inserting $g = 1/3$, $f_{\mathrm{a}} = f_{\mathrm{b}} = 0.5$, and $\varepsilon_{\mathrm{a}} = \varepsilon_{\mathrm{b}} = 1$. Then the effective dielectric function has to take the value of vacuum, which is possible only by allowing a positive sign for the square root.

The spectroscopic properties of liquids are usually measured by a reflectometer [63], which is based on the use of a prism as a probe. The information about the liquid is obtained from a thin layer at the prism–liquid interface in the region of an evanescent wave. For example, the islands of oil drops on the surface of water can be treated as a Bruggeman effective medium. Nevertheless, reflectometric study yields information only from a thin layer at the probe–sample interface. In the measurement of the effective reflectivity of a random nanostructure, the effective dielectric function at the probe–sample interface may differentiate from the sample average, which leads to an erroneous result. Therefore, Gehr et al. [64] suggested a measurement over the total volume based on the use of an interferometer and found good agreement between predicted values and measured data.

Layered Nanostructures

In Fig. 3.2c, we present a two-phase layered nanostructure, built by alternating layers of subwavelength thicknesses. Layered nanostructures are assumed to have a wide range of optoelectronic applications from all-optical switching, modulating, and computing devices to quantum well lasers.

For TE-polarized light (the electric field is parallel to the layers), the tangential component of the electric field E_{TE} is continuous at the boundary between the layers. The electric field is spatially distributed uniformly between the constituents a and b with volume fractions f_a and f_b. The average electric displacement inside the nanostructure can be expressed as follows:

$$D_\mathrm{TE}(\omega) = f_\mathrm{a} D_\mathrm{TE}^{(\mathrm{a})}(\omega) + f_\mathrm{b} D_\mathrm{TE}^{(\mathrm{b})}(\omega) = [f_\mathrm{a}\varepsilon_\mathrm{a}(\omega) + f_\mathrm{b}\varepsilon_\mathrm{b}(\omega)]E_\mathrm{TE}(\omega), \quad (3.66)$$

since E_TE is continuous over the structure. On the other hand, the average electric displacement is related to the effective dielectric function by the following relation:

$$D_\mathrm{TE}(\omega) = \varepsilon_\mathrm{eff}(\omega) E_\mathrm{TE}(\omega). \quad (3.67)$$

By combining (3.66) and (3.67), it is possible to obtain the expression for the effective dielectric function for TE-polarized light:

$$\varepsilon_\mathrm{eff}(\omega) = f_\mathrm{a}\varepsilon_\mathrm{a}(\omega) + f_\mathrm{b}\varepsilon_\mathrm{b}(\omega). \quad (3.68)$$

For TM-polarized light (the electric field is perpendicular to the layers), the normal component of the electric displacement D_TM is continuous at the boundary. The average electric field is

$$E_\mathrm{TM}(\omega) = f_\mathrm{a} E_\mathrm{TM}^{(\mathrm{a})}(\omega) + f_\mathrm{b} E_\mathrm{TM}^{(\mathrm{b})}(\omega) = \left[\frac{f_\mathrm{a}}{\varepsilon_\mathrm{a}(\omega)} + \frac{f_\mathrm{b}}{\varepsilon_\mathrm{b}(\omega)}\right] D_\mathrm{TM}(\omega), \quad (3.69)$$

where D_TM is continuous. It should be noted that the electric field is distributed nonuniformly between the constituents. The effective dielectric function for TM-polarized light is given by

$$\frac{1}{\varepsilon_\mathrm{eff}(\omega)} = \frac{f_\mathrm{a}}{\varepsilon_\mathrm{a}(\omega)} + \frac{f_\mathrm{b}}{\varepsilon_\mathrm{b}(\omega)}. \quad (3.70)$$

Finally, it is worth mentioning that linear optical properties are independent of the thicknesses of individual layers. Hence, only the volume fraction of different constituents matters. We emphasize that the linear optical response of a layered nanostructure is anisotropic, thus yielding a form of birefringence: the refractive index of the medium is different for TE- and TM-polarized light.

Asymptotic Behavior of Linear Susceptibility for Nanostructures

We have shown in (3.56) that the asymptotic behavior of the effective linear susceptibility relative to each material constituting the nanostructure can be expressed as follows:

$$\chi_{\mathrm{eff},i}^{(1)}(\omega) = -\frac{\omega_{\mathrm{p}i}^2}{4\pi}\frac{1}{\omega^2} + o(\omega^{-2}), \quad (3.71)$$

where the index i defines the material we are referring to, host and inclusion in the MG and Bruggemann cases, and layer 1 and layer 2 in the layered nanostructure case, and $\omega_{\mathrm{p}i}$ is the corresponding plasma frequency. Considering for both phases as well as for the effective medium, the usual correspondence between the dielectric function and the susceptibility presented in (3.6), the asymptotic behavior of the effective susceptibilities of the systems considered can be found by direct substitution of the asymptotic behavior (3.71) in the expressions obtained for the effective dielectric functions. We find that in all cases analyzed – MG, Bruggemann, and layered nanostructures with TE- and TM-polarized light – the asymptotic falloff of the effective susceptibility is

$$\chi_{\mathrm{eff}}^{(1)}(\omega) = -\frac{[f_{\mathrm{a}}\omega_{\mathrm{pa}}^2 + f_{\mathrm{b}}\omega_{\mathrm{pb}}^2]}{4\pi}\frac{1}{\omega^{-2}} + o(\omega^{-2}). \tag{3.72}$$

We emphasize that asymptotic behavior (3.72) is the volume average of the asymptotic behaviors of the two materials, regardless of the topology of the system.

4 Kramers-Kronig Relations and Sum Rules in Linear Optics

4.1 Introductory Remarks

Building upon the physical framework developed in the previous chapters, we derive in all generality the Kramers-Kronig relations and the related sum rules for linear optical functions. The dispersion relations and the sum rules are then generalized also for the analysis of the linear optical properties of the most typical nanostructures. A very complete theoretical treatment of these issues can be found in [34].

In linear optical spectroscopy, K-K analysis has two typical functions depending on whether the measurement is based on the transmission or reflection of light. In the former case, usually the imaginary part is measured and the real part is obtained by a K-K relation, while in the latter case, the intensity is measured and the phase is calculated by the appropriate K-K relation. However, when using an ellipsometer, it is possible to obtain experimental information from the complex function obtained by the measurement. In such a case, K-K relations can be used to test the self-consistency of the measured real and imaginary parts of the data.

We emphasize that linear optical spectroscopy is probably the most thoroughly exploited tool in optical materials research. For instance, spectrophotometers that utilize the Beer-Lambert law of light absorption have been commercially available for a long time.

4.2 The Principle of Causality

Causality is one of the fundamental principles in physics. It states that the effect cannot precede the cause. Nussenzveig [2] presents an interesting discussion about different definitions of causality. For optical purposes, the most useful concept is relativistic causality, which states that no signal can propagate faster than the speed of light in vacuum. The connection between causality and a scattering matrix (S-matrix) was studied widely in the 1950s [65–67]. With the aid of the S-matrix, it is possible to transform a system from the initial state to the final state. Causality indicates that no scattered wave can exist before the incident wave has reached the scattering center, whose size is assumed to be finite. Causality implies that the general S-matrix has an analytic (also referred to as holomorphic) continuation in the upper complex energy plane and complex poles (= singular points) are located in the lower half of the complex energy plane. In optical physics, with the aid of analytic continuation of the refractive index to the upper half of a complex angular frequency plane, Kramers [4] proved that the principle of relativistic causality allows the calculation of the real refractive index of a medium from the absorption spectrum. Kronig [5] showed that the existence of a dispersion relation is a sufficient and a necessary condition for strict causality to hold. However, he made assumptions on the holomorphic behavior of the investigated function. A rigorous statement of equivalence between causality and the existence of dispersion relations, and so of the existence of a strict connection between the mathematical properties of the functions describing the physics in the domains of time and frequency, is provided by the Titchmarsch theorem [2].

4.3 Titchmarsch's Theorem and Kramers-Kronig Relations

Titchmarsch's theorem connects, within fairly loose hypotheses, the causality of a function $a(t)$ to the analytical properties of its Fourier transform $a(\omega) = \mathrm{F}[a(t)]$

Theorem 1. (Titchmarsch)

The three statements 1, 2, and 3 are mathematically equivalent:

1. $a(t) = 0$ if $t \leq 0$ and $a(t) \in \mathrm{L}^2$.
2. $a(\omega) = \mathrm{F}[a(t)] \in \mathrm{L}^2$ if $\omega \in \mathbb{R}$ and if

$$a(\omega) = \lim_{\omega' \to 0} a(\omega + \mathrm{i}\omega'),$$

   then $a(\omega + \mathrm{i}\omega')$ is holomorphic if $\omega' > 0$.
3. Hilbert transforms [39] connect the real and imaginary parts of $a(\omega)$ as follows:

$$\mathrm{Re}\{a(\omega)\} = \frac{1}{\pi} \mathrm{P} \int_{-\infty}^{\infty} \frac{\mathrm{Im}\{a(\omega')\}}{\omega' - \omega} \mathrm{d}\omega',$$

$$\mathrm{Im}\{a(\omega)\} = -\frac{1}{\pi} \mathrm{P} \int_{-\infty}^{\infty} \frac{\mathrm{Re}\{a(\omega')\}}{\omega' - \omega} \mathrm{d}\omega'.$$

Thus, the causality of $a(t)$, together with its property of being a function belonging to the space of the square-integrable functions L^2, implies that its Fourier transform $a(\omega)$ is analytic in the upper complex ω-plane and that the real and imaginary parts of $a(\omega)$ are not independent but are connected by nonlocal, integral relations termed dispersion relations.

Under the reasonable assumption that all tensorial components of the linear Green function belong to the L^2 space, we deduce that Hilbert transforms connect the real and imaginary parts of the tensorial components of linear susceptibility. Considering that the components of polarization $\boldsymbol{P}^{(1)}(t)$ are real functions, we can deduce that

$$\chi_{ij}^{(1)}(-\omega) = \left[\chi_{ij}^{(1)}(\omega)\right]^*, \tag{4.1}$$

where (*) denotes the complex conjugate. Equation (4.1) implies that for every $\omega \in \mathbb{R}$, the following relations hold:

$$\mathrm{Re}\left\{\chi_{ij}^{(1)}(\omega)\right\} = \mathrm{Re}\left\{\chi_{ij}^{(1)}(-\omega)\right\}, \tag{4.2}$$

$$\mathrm{Im}\left\{\chi_{ij}^{(1)}(\omega)\right\} = -\mathrm{Im}\left\{\chi_{ij}^{(1)}(-\omega)\right\}. \tag{4.3}$$

Taking advantage of these expressions, we can finally write the K-K relations [4,5]

$$\mathrm{Re}\left\{\chi_{ij}^{(1)}(\omega)\right\} = \frac{2}{\pi}\mathrm{P}\int_0^\infty \frac{\omega'\mathrm{Im}\left\{\chi_{ij}^{(1)}(\omega')\right\}}{\omega'^2 - \omega^2}\mathrm{d}\omega', \tag{4.4}$$

$$\mathrm{Im}\left\{\chi_{ij}^{(1)}(\omega)\right\} = -\frac{2\omega}{\pi}\mathrm{P}\int_0^\infty \frac{\mathrm{Re}\left\{\chi_{ij}^{(1)}(\omega')\right\}}{\omega'^2 - \omega^2}\mathrm{d}\omega'. \tag{4.5}$$

We note that the static susceptibility can be obtained from (4.4) simply by setting $\omega = 0$.

The relevance of K-K relations in physics goes beyond the purely conceptual sphere. The real and imaginary parts of susceptibility are related to qualitatively different phenomena, light dispersion and light absorption, respectively, whose measurement requires different experimental setups and instruments. We underline that since every function that models the linear susceptibility of a material, as well as any set of experimental data of the real and imaginary parts of linear susceptibility, must respect the K-K relations, these also constitute a fundamental test of self-consistency [10, 12–15, 38].

4.3.1 Kramers-Kronig Relations for Conductors

In the special case of conductors, the dispersion relations have to be modified [68–70] because of the presence of a pole in the linear susceptibility for $\omega = 0$, and so the first part of statement 2 in the Titchmarsch theorem is not satisfied [12].

It is possible in general to express susceptibility as a function of linear conductivity [35, 38]:

$$\chi_{ij}^{(1)}(\omega) = \mathrm{i}\frac{\sigma_{ij}^{(1)}(\omega)}{\omega} = \mathrm{i}\frac{\mathrm{Re}\left\{\sigma_{ij}^{(1)}(\omega)\right\}}{\omega} - \frac{\mathrm{Im}\left\{\sigma_{ij}^{(1)}(\omega)\right\}}{\omega}, \tag{4.6}$$

where the linear conductivity tensor describes the physically indistinguishable effects of polarization and current. The conductors, by definition, have a nonvanishing real conductivity for static electric fields. In consequence, the imaginary part of the linear susceptibility has a pole for $\omega = 0$, i.e. the following relation holds for low values of ω:

$$\mathrm{Im}\left\{\chi_{ij}^{(1)}(\omega)\right\} \approx \frac{\sigma_{ij}^{(1)}(0)}{\omega}. \tag{4.7}$$

As a consequence, when computing the Hilbert transform of linear susceptibility, there is a boundary contribution from the pole at the origin. This determines a change in the integration path in the complex ω-plane from the usual case [15], so that the K-K relation for conductors can be written as

$$\mathrm{Re}\left\{\chi_{ij}^{(1)}(\omega)\right\} = \frac{2}{\pi}\mathrm{P}\int_0^\infty \frac{\omega'\mathrm{Im}\left\{\chi_{ij}^{(1)}(\omega')\right\}}{\omega'^2 - \omega^2}\mathrm{d}\omega', \tag{4.8}$$

$$\mathrm{Im}\left\{\chi_{ij}^{(1)}(\omega)\right\} - \frac{\sigma_{ij}^{(1)}(0)}{\omega} = -\frac{2\omega}{\pi}\mathrm{P}\int_0^\infty \frac{\mathrm{Re}\left\{\chi_{ij}^{(1)}(\omega')\right\}}{\omega'^2 - \omega^2}\mathrm{d}\omega'. \tag{4.9}$$

We observe that if the static conductance is vanishing, these modified K-K relations coincide with the usual dispersion relations presented in (4.4) and (4.5).

4.3.2 Kramers-Kronig Relations for the Effective Susceptibility of Nanostructures

The linear optical properties of nanostructures can be described with the effective dielectric function of the structure. Expressions for the effective dielectric function of different models, including MG, Bruggeman, and layered nanostructures, have been presented in the previous chapter.

Since the consideration of the local field effects provides contributions to the Hamiltonian of the system but does not alter the causal nature of the Green function of the system, $\chi_{\mathrm{eff}}^{(1)}(\omega)$ is holomorphic in the upper complex ω-plane. Moreover, it has been proved by direct calculation that the effective dielectric function $\chi_{\mathrm{eff}}^{(1)}(\omega)$ fulfills the symmetry property $\chi_{\mathrm{eff}}^{(1)}(-\omega) = \left[\chi_{\mathrm{eff}}^{(1)}(-\omega)\right]^*$ in the cases of MG [62], Bruggeman [71], and layered [72] nanostructures. These properties and the asymptotic equivalence between $\chi^{(1)}(\omega)$ and $\chi_{\mathrm{eff}}^{(1)}(\omega)$ shown in (3.72) imply that the latter function obeys the same K-K relations as the former function:

$$\mathrm{Re}\{\chi_{\mathrm{eff}}^{(1)}(\omega)\} = \frac{2}{\pi}\mathrm{P}\int_0^\infty \frac{\omega'\mathrm{Im}\{\chi_{\mathrm{eff}}^{(1)}(\omega')\}}{\omega'^2-\omega^2}\mathrm{d}\omega', \tag{4.10}$$

$$\mathrm{Im}\{\chi_{\mathrm{eff}}^{(1)}(\omega)\} = -\frac{2\omega}{\pi}\mathrm{P}\int_0^\infty \frac{\mathrm{Re}\{\chi_{\mathrm{eff}}^{(1)}(\omega')\}}{\omega'^2-\omega^2}\mathrm{d}\omega'. \tag{4.11}$$

Therefore, the results obtained in this chapter are still valid in the more general and realistic context where local field effects are considered.

Following the work by Stroud [73], we can define the average linear susceptibility function of a nanostructure by the form

$$\chi_{\mathrm{av}}^{(1)}(\omega) = f_{\mathrm{a}}\chi_{\mathrm{a}}^{(1)}(\omega) + f_{\mathrm{b}}\chi_{\mathrm{a}}^{(1)}(\omega). \tag{4.12}$$

It is easy to find that the dominant term in the asymptotic behavior of the function $\chi_{\mathrm{av}}^{(1)}(\omega)$ is the same as for the function $\chi_{\mathrm{eff}}^{(1)}(\omega)$. Considering the parity properties of the real part of the susceptibility functions, we can then derive the asymptotic falloff of the difference between the average and effective susceptibility:

$$\chi_{\mathrm{eff}}^{(1)}(\omega) - \chi_{\mathrm{av}}^{(1)}(\omega) \sim \frac{\psi}{\omega^4} + O(\omega^{-4}), \tag{4.13}$$

where ψ is a constant. The functions $\chi_{\mathrm{eff}}^{(1)}(\omega) - \chi_{\mathrm{av}}^{(1)}(\omega)$ and $\omega^{2\alpha}[\chi_{\mathrm{eff}}^{(1)}(\omega) - \chi_{\mathrm{av}}^{(1)}(\omega)]$ are holomorphic in the upper complex plane of the variable ω and have an asymptotic decrease strictly faster than ω^{-1}, therefore the following independent K-K relations connect the real and imaginary parts:

$$\omega^{2\alpha}\mathrm{Re}\{\chi_{\mathrm{eff}}^{(1)}(\omega) - \chi_{\mathrm{av}}^{(1)}(\omega)\} = \frac{2}{\pi}\mathrm{P}\int_0^\infty \frac{\omega'^{2\alpha+1}\mathrm{Im}\{\chi_{\mathrm{eff}}^{(1)}(\omega') - \chi_{\mathrm{av}}^{(1)}(\omega')\}}{\omega'^2-\omega^2}\mathrm{d}\omega', \tag{4.14}$$

$$\omega^{2\alpha-1}\mathrm{Im}\{\chi_{\mathrm{eff}}^{(1)}(\omega) - \chi_{\mathrm{av}}^{(1)}(\omega)\} = -\frac{2}{\pi}\mathrm{P}\int_0^\infty \frac{\omega'^{2\alpha}\mathrm{Re}\{\chi_{\mathrm{eff}}^{(1)}(\omega') - \chi_{\mathrm{av}}^{(1)}(\omega')\}}{\omega'^2-\omega^2}\mathrm{d}\omega', \tag{4.15}$$

with $\alpha = 0, 1$. We note that when we consider suitable functions with a faster asymptotic decrease, it is generally possible to derive a larger set of independent K-K relations.

4.4 Superconvergence Theorem and Sum Rules

From the K-K relations and knowledge of the asymptotic behavior of the susceptibility function, it is possible to deduce the value of the zero-degree moment of the real part and the value of the first moment of the imaginary part of the susceptibility. This result can be obtained by taking advantage of the superconvergence theorem [9, 74]

Theorem 2. (superconvergence)

If

$$g(y) = \mathrm{P} \int_0^\infty \frac{f(x)}{y^2 - x^2} \mathrm{d}x,$$

where

1. $f(x)$ is continuously differentiable,
2. $f(x) = O\left[(x \ln x)^{-1}\right]$,

then for $y \gg x$, the following asymptotic expansion holds:

$$g(y) = \frac{1}{y^2} \int_0^\infty f(x)\, \mathrm{d}x + O\left(y^{-2}\right).$$

The theorem can also be written in the form

$$\int_0^\infty f(x)\, \mathrm{d}x = \lim_{y \to \infty} \left[y^2 g(y)\right]. \tag{4.16}$$

We then consider the K-K relation (4.4), we replace $x = \omega'$, $y = \omega$, and we set $f(x)$ and $g(y)$ equal to the general tensorial component of the imaginary and real parts of the linear susceptibility, respectively. The discussion presented in this chapter shows that the asymptotic decrease for all tensorial components of the imaginary part is strictly faster than ω^{-2}. Therefore, by applying the superconvergence theorem [9, 74] to the K-K relation (4.4), we conclude that

$$\int_0^\infty \omega' \mathrm{Im}\left\{\chi_{ij}^{(1)}(\omega')\right\} \mathrm{d}\omega' = \lim_{\omega \to \infty} \left(-\frac{\pi}{2}\omega^2 \mathrm{Re}\left\{\chi_{ij}^{(1)}(\omega)\right\}\right) = \frac{\omega_{\mathrm{p}}^2}{8}\delta_{ij}, \tag{4.17}$$

where we have considered the general result for the asymptotic behavior of the real part of the susceptibility presented in (3.44). This law is commonly referred to as Thomas-Reiche-Kuhn (TRK) or f sum rule [6–8]. Considering that the quantity under integral is proportional to the absorption of the material under examination, we may see that the total absorption over all the spectrum is proportional to the total electronic density. It is possible to obtain the following sum rule by applying the superconvergence theorem to the second dispersion relation (4.5):

$$\int_0^\infty \mathrm{Re}\left\{\chi_{ij}^{(1)}(\omega')\right\} \mathrm{d}\omega' = \lim_{\omega \to \infty} \left(\frac{\pi}{2}\omega \mathrm{Im}\left\{\chi_{ij}^{(1)}(\omega)\right\}\right) = 0. \tag{4.18}$$

This sum rule implies that the average value of the real part of the dielectric function over all the spectrum is 1.

Sum rules have long constituted fundamental tools in the investigation of light–matter interaction phenomena in condensed matter, gases, molecules, and liquids. They provide constraints for checking the self-consistency of experimental or model-generated data [10, 12–15, 38].

4.5 Sum Rules for Conductors

Applying the superconvergence theorem in the usual way to the K-K relations for the conductors presented in (4.8) and (4.9), we find that while the TRK sum rule is still obeyed, the boundary term contained in the second K-K relation changes the sum rule for the real part of the susceptibility:

$$\begin{aligned}\int_0^\infty \mathrm{Re}\left\{\chi_{ij}^{(1)}(\omega')\right\} \mathrm{d}\omega' &= \lim_{\omega \to \infty} \left[\frac{\pi}{2}\omega \left(\mathrm{Im}\left\{\chi_{ij}^{(1)}(\omega)\right\} - \frac{\sigma_{ij}^{(1)}(0)}{\omega}\right)\right] \\ &= -\frac{\pi}{2}\sigma_{ij}^{(1)}(0)\end{aligned} \tag{4.19}$$

Therefore, in the case of conductors, the average value of the dielectric function is not equal to 1 [68].

We point out that the verification of linear sum rules from experimental data is usually hard to obtain because of the critical contributions made by the out-of-range asymptotic part of the real or imaginary parts of the susceptibility under examination [12, 75, 76]. However, in the case of linear optics, information about the response of the material to very high frequency radiation can be obtained using synchrotron radiation [12].

In particular, we wish to emphasize that the integral properties obtained for the linear susceptibility that we have derived by adopting a quantum mechanical description of the matter have a one-to-one correspondence with those that can be obtained by suitably modeling the matter with a simple classical mechanical picture. In such a picture, each electron is independent, is bound within a given volume by a harmonic potential, oscillates under the influence of the external oscillating electric field, and is slowed down by linear friction. In the special case of conductors, the potential is assumed to be vanishing. The traditionally termed Drude-Lorentz oscillator model [12, 14, 15, 17, 35, 38, 77] shows that the integral properties we have analyzed in this chapter are fundamental properties of a wide class of systems that transcend how the system is modeled, since they descend from assuming causality in the interaction processes.

4.5.1 Sum Rules for the Linear Effective Susceptibilities of Nanostructures

We have previously presented K-K relations (4.10) and (4.11) for the effective susceptibility of nanostructures for all the investigated topologies. Therefore, by applying the superconvergence theorem to a pair of K-K relations and taking into account the general asymptotic behavior presented in (3.72), it is possible to derive sum rules for the real and imaginary parts of the effective susceptibility. The corresponding TRK sum rule for the effective susceptibility is of the form

$$\int_0^\infty \omega' \mathrm{Im}\{\chi_{\mathrm{eff}}^{(1)}(\omega')\} \mathrm{d}\omega' = \frac{1}{8}\left[f_\mathrm{i}\omega_{\mathrm{pi}}^2 + (1-f_\mathrm{i})\omega_{\mathrm{ph}}^2\right]. \tag{4.20}$$

It is worth noting that the high-frequency behavior of the composite transforms the TRK sum rule according to the volume fractions and plasma frequencies of the constituents, while the average sum rule is valid for all topologies:

$$\int_0^\infty \mathrm{Re}\{\chi_{\mathrm{eff}}^{(1)}(\omega')\} \mathrm{d}\omega' = 0, \tag{4.21}$$

which is totally analogous to that presented in (4.18).

Moreover, if we consider the set of K-K relations for the real and imaginary parts of the function $\chi_{\mathrm{eff}}^{(1)}(\omega) - \chi_{\mathrm{av}}^{(1)}(\omega)$ presented in (4.14) and (4.15), by applying the superconvergence theorem to the equations, we obtain the following sum rules:

$$\int_0^\infty \omega'^{2\alpha} \mathrm{Re}\{\chi_{\mathrm{eff}}^{(1)}(\omega') - \chi_{\mathrm{av}}^{(1)}(\omega')\} \mathrm{d}\omega' = 0, \qquad \alpha = 0, 1 \tag{4.22}$$

$$\int_0^\infty \omega'^{2\alpha+1} \mathrm{Im}\{\chi_{\mathrm{eff}}^{(1)}(\omega') - \chi_{\mathrm{av}}^{(1)}(\omega')\} \mathrm{d}\omega' = \begin{cases} 0 & , \quad \alpha = 0 \\ -\frac{\pi}{2}\psi & , \quad \alpha = 1 \end{cases}. \tag{4.23}$$

The sum rule (4.23) for $\alpha = 1$ was first given by Stroud [73] in the case of an MG composite, for which

$$\psi = -\frac{1}{3} f_\mathrm{a} f_\mathrm{b} [\omega_{\mathrm{pb}}^2 - \omega_{\mathrm{pa}}^2]^2. \tag{4.24}$$

4.6 Integral Properties of Optical Constants

The linear susceptibility function describes, at a fundamental level, the connection between the microscopic dynamics of the system under consideration and its linear optical properties. Nevertheless, it is experimentally much easier to measure other quantities that are more directly related to the behavior of light influenced by its interaction with matter. The most commonly used optical constants are the complex index of refraction $N(\omega) = \eta(\omega) + \mathrm{i}\kappa(\omega)$ and the complex reflectivity at normal incidence $r(\omega)$. It has been been shown that relevant integral relations can also be established for these quantities.

Hence, the K-K relations and sum rules find wider experimental application in the analysis of the data for these optical constants. In this section, we consider the simplifying assumption of isotropic matter, and we briefly present the integral properties of the complex index of refraction $N(\omega)$. We show that that they are analogous to those presented for linear susceptibility. We will not treat the experimentally relevant case of reflectivity at normal incidence $r(\omega)$ since it requires a slightly more complex and cumbersome mathematical treatment; a very detailed treatment of integral relations for $r(\omega)$ can be found in [15].

4.6.1 Integral Properties of the Index of Refraction

We observe that the complex index of refraction $N(\omega)$ presented in (3.16) is the square root of a function that is is holomorphic in the upper complex ω-plane. Following the procedure proposed in [2], it is possible to determine that $N(\omega)$ itself is holomorphic, has no branching points in the upper complex ω-plane, and that the usual crossing relation $N(\omega) = [N(-\omega)]^*$ holds. Moreover, it is simple to prove that the function $N(\omega) - 1$ decreases asymptotically as:

$$N(\omega) - 1 \sim -\frac{\omega_{\mathrm{p}}^2}{2\omega^2}. \tag{4.25}$$

We conclude that we can write the following K-K relations connecting the real and imaginary parts of the refractive index:

$$\eta(\omega) - 1 = \frac{2}{\pi}\mathrm{P}\int_0^\infty \frac{\omega'\kappa(\omega')}{\omega'^2 - \omega^2}\mathrm{d}\omega', \tag{4.26}$$

$$\kappa(\omega) = -\frac{2\omega}{\pi}\mathrm{P}\int_0^\infty \frac{\eta(\omega') - 1}{\omega'^2 - \omega^2}\mathrm{d}\omega'. \tag{4.27}$$

We wish to emphasize that this is the original formulation of K-K relations proposed by Kramers [4]. Moreover, it should be noted that these dispersion relations (4.26) are valid both in the case of conductors and nonconductors; we refer to [78] for a thorough discussion of this issue. The usual protocol in linear optical spectroscopy is the measurement of light transmission through a solid or liquid sample. Since the thickness d of the sample is known and the (dual beam) spectrophotometer provides the wavelength-dependent optical density, the absorption coefficient can be resolved by the Beer-Lambert intensity law, $I = I_0 \exp(-\alpha(\omega)d)$. There is a simple relation between the extinction coefficient $\kappa(\omega)$ and the absorption coefficient $\alpha(\omega)$ of the medium, i.e. $\kappa(\omega) = c\alpha(\omega)/2\omega$, which implies that it is possible to exploit (4.26) to find the refractive index change. As an example of such data inversion, we show in Fig. 4.1 the extinction curves and corresponding refractive index changes of a KBr crystal containing F and M color centers. The curves were obtained by

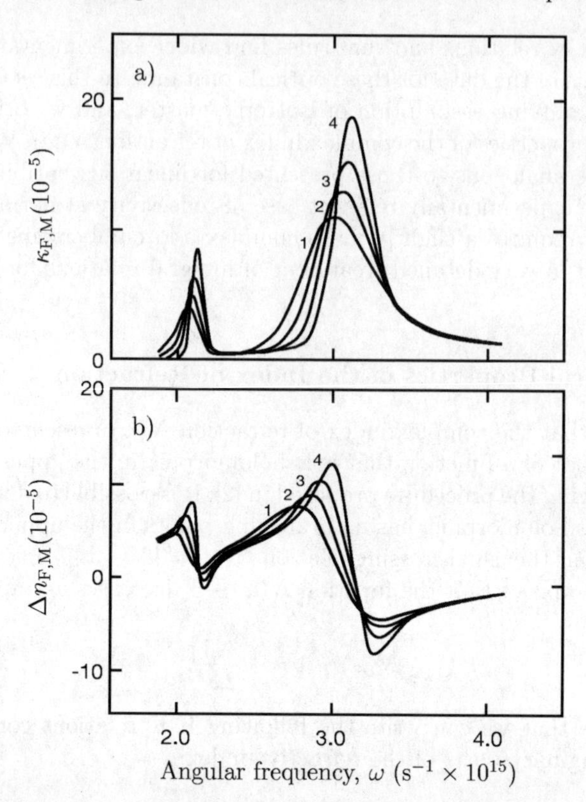

Fig. 4.1. (**a**) Extinction curves and (**b**) refractive index changes of F and M color centers in a KBr crystal at different temperatures. Note in (**b**) that for certain temperatures there are three angular frequency values for which $\Delta\eta_{F,M} = 0$. The curves are 1–250 K, 2–170 K, 3–100 K, and 4–35 K. Reproduced from [80]

measuring light transmission as a function of wavelength and temperature. In Figs. 4.2–4.3, we show the result of other applications of K-K relations to the experimental data of $\kappa(\omega)$. From the measurement of the absorbance only, the authors have been able to deduce all the information needed to reconstruct the real and imaginary parts of the dielectric function [79]. If we apply the superconvergence theorem to (4.26) and consider the asymptotic behavior presented in (4.25), we obtain

$$\int_0^\infty \omega' \kappa(\omega') \, d\omega' = \lim_{\omega \to \infty} \left\{ -\frac{\pi}{2} \omega^2 \left[\eta(\omega) - 1 \right] \right\} = \frac{\pi \omega_p^2}{4}, \qquad (4.28)$$

which is equivalent to the TRK sum rule [6–8] presented in (4.17). Following the same procedure for the second dispersion relation (4.27), we obtain the Altarelli-Dexter-Nussenzveig-Smith (ADNS) sum rule [9]:

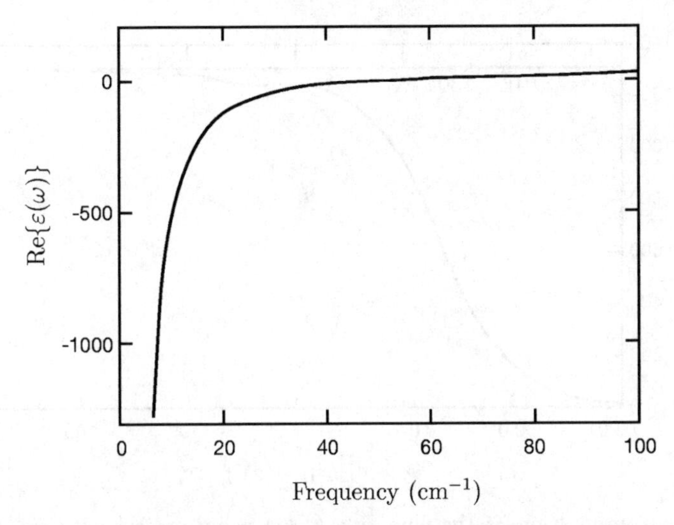

Fig. 4.2. Real part of the dielectric function $\varepsilon(\omega)$ obtained from K-K analysis of the absorbance of $La_{1.87}Sr_{0.13}CuO_4$. Reproduced from [79]

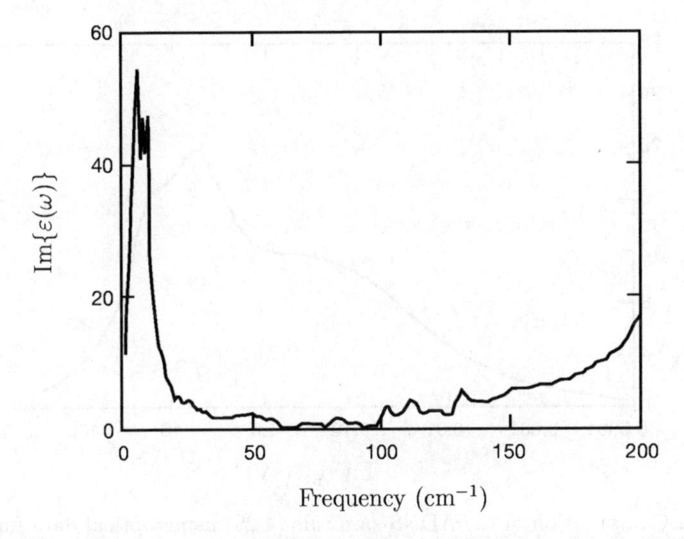

Fig. 4.3. Imaginary part of the dielectric function $\varepsilon(\omega)$ obtained from K-K analysis of the absorbance of $La_{1.87}Sr_{0.13}CuO_4$. Reproduced from [79]

$$\int_0^\infty [\eta(\omega') - 1]\mathrm{d}\omega' = \lim_{\omega\to\infty}\left[\frac{\pi}{2}\omega\kappa(\omega)\right] = 0. \tag{4.29}$$

We underline that this sum rule is not wholly equivalent to the sum rule (4.18) obtained for the imaginary part of the susceptibility. The reason is that if we compute the sum rule for the imaginary part of the susceptibility function in

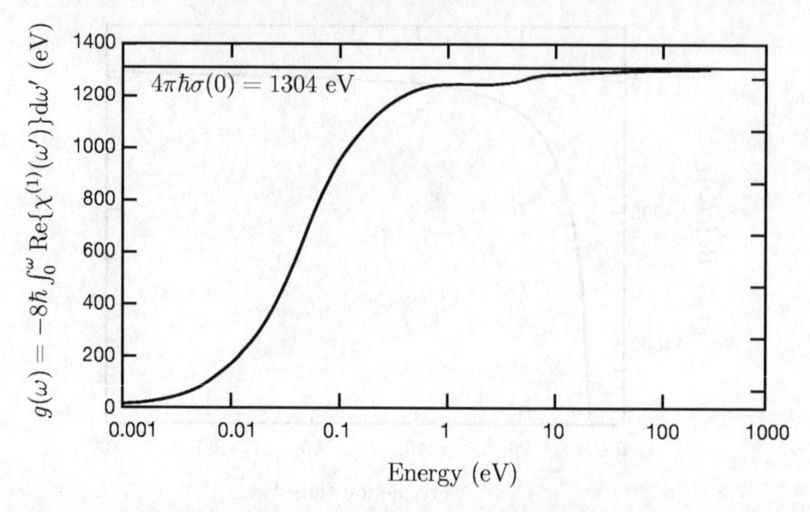

Fig. 4.4. Computation of the sum rule (4.18) using optical data for molybdenum. Note that, instead of $\mathrm{Re}\{\chi^{(1)}(\omega)\}$, the equivalent quantity $1 - \mathrm{Re}\{\varepsilon(\omega)\} = -4\pi\mathrm{Re}\{\chi^{(1)}(\omega)\}$ is considered in the integration. Reproduced from [81]

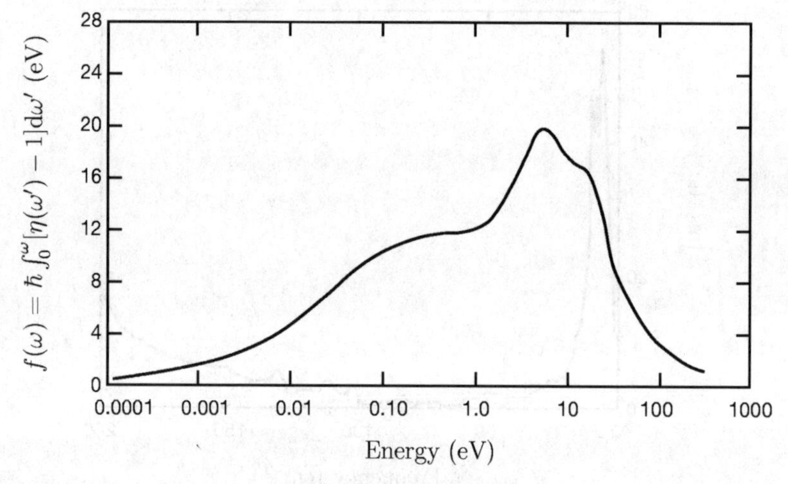

Fig. 4.5. Computation of the ADNS sum rule (4.29) using optical data for molybdenum. Note that a factor $\hbar$ appears in this figure. Reproduced from [81]

the case of conductors, we can obtain a measure of the static conductance as expressed in (4.19); the result (4.29) is valid in both cases of conducting and nonconducting matter. We propose that the reason why the consideration of susceptibility permits us to discern between conductors and nonconductors is that susceptibility is a more fundamental physical quantity, since it is directly related to the dynamics of the system. In Figs. 4.4 and 4.5, we show the results obtained for the integration of the sum rule for the real part of the susceptibility presented in expression (4.18) and the sum rule for $[\eta(\omega) - 1]$ presented in (4.29) for optical data for molybdenum [81], respectively. While in the former case, the integral converges asymptotically to a positive quantity, which corresponds to the actual conductivity as predicted by (4.18), in the latter case, the integral converges to zero, as predicted by the ADNS sum rule (4.29).

We remark that in practical data analysis, the usual problem related to the utilization of K-K relations, of alternative methods based on Fourier transforms [82], and of sum rules is the requirement of extrapolating data beyond the measured spectral range in order to achieve a wider domain of integration. Nowadays, different software packages are available for spectroscopic devices that make use of K-K analysis. However, it has been shown that the utilization of such software may lead to qualitatively different results when using identical input data [83]. Such problems can be eased by using more advanced inversion methods, such as multiply subtractive Kramers-Kronig relations and the maximum entropy methods, which will be introduced later in the book. Finally, we wish to remark that K-K relations and sum rules are valid even in the frame of negative refractive index media [84, 85]. Such media have drawn much interest recently [86, 87].

4.6.2 Kramers-Kronig Relations in Linear Reflectance Spectroscopy

The estimation of the optical constants of a sample is commonly obtained by measuring the reflectance at normal incidence and using dispersion relations in the phase retrieval procedure. Measurement of normal reflectance is especially well suited for the determination of optical constants in regions of high absorbance [88]. The first significant experimental applications of normal reflectivity measurements for the determination of the optical constants of semiconductors were carried out in 1962 by Ehrenreich and Philipp [89,90]. Since then, the method has become a standard spectroscopic technique [10, 91]. At normal incidence, the complex reflectivity $r(\omega)$ of light at the boundary between sample and vacuum is related to the complex refractive index of the sample, according to Fresnel's equations (3.24) and (3.28):

$$r(\omega) = \frac{N(\omega) - 1}{N(\omega) + 1} = \frac{\eta(\omega) - 1 + \mathrm{i}\kappa(\omega)}{\eta(\omega) + 1 + \mathrm{i}\kappa(\omega)}. \tag{4.30}$$

There is an analytic continuation of function $r(\omega)$ from the real axis into the upper half of the complex angular frequency ω-plane [88]. In the case of a dielectric, the real axis belongs to the domain of holomorphicity, while in the case of conductors, there is a branch point at $\omega = 0$. The effect of the branch point on a dispersion relation for a phase spectrum has been discussed by Nash et al. [92] and Lee and Sindoni [93]. It is possible to establish a symmetry relation for complex reflectivity with linearly polarized light:

$$r(-\omega^*) = [r(\omega)]^*. \tag{4.31}$$

For circularly and elliptically polarized light, the reflectivity has a lower symmetry after the mixing of left- and right-hand modes [94]. The quantity obtained from the reflectance measurement is usually the amplitude of the reflected beam, which is the square of the modulus of the complex reflectivity:

$$R(\omega) = |r(\omega)|^2 = r(\omega)[r(\omega)]^*. \tag{4.32}$$

Complex reflectivity can be expressed in polar coordinates as $r(\omega) = |r(\omega)| \exp[\mathrm{i}\theta(\omega)]$, so that we obtain

$$\ln r(\omega) = \ln |r(\omega)| + \mathrm{i}\theta(\omega). \tag{4.33}$$

Unfortunately, the function diverges logarithmically at the limit $|\omega| \to \infty$ and is not square-integrable [95]. Divergence of the integrals eliminates the possibility of deriving analogous dispersion relations for complex reflectivity as for optical constants [96]. Another important requirement for complex reflectivity is the lack of zeros [88], which are branch points of a logarithm. However, there are several possibilities for avoiding the divergence of the integral. On one hand, the function [97]

$$F(\omega) = \frac{\ln r(\omega)}{\omega^2 - \omega'^2} \tag{4.34}$$

gives the well-known relation for the phase of complex reflectivity, as follows:

$$\theta(\omega) = -\frac{2\omega}{\pi} \mathrm{P} \int_0^\infty \frac{\ln |r(\omega')|}{\omega'^2 - \omega^2} \mathrm{d}\omega'. \tag{4.35}$$

On the other hand, the function [97]

$$G(\omega) = \ln r(\omega) \left[\frac{1}{\omega - \omega_2} - \frac{1}{\omega - \omega_2} \right] \tag{4.36}$$

gives the dispersion relation for the amplitude of complex reflectivity

$$\ln |r(\omega_1)| - \ln |r(\omega_2)| = \frac{2}{\pi} \mathrm{P} \int_0^\infty \omega' \theta(\omega') \left[\frac{1}{\omega'^2 - \omega_1^2} - \frac{1}{\omega'^2 - \omega_2^2} \right] \mathrm{d}\omega'. \tag{4.37}$$

Although the expression (4.37) seems to be nothing but a subtracted K-K relation, it is not. Both functions $\ln |r(\omega_1)|$ and $\ln |r(\omega_2)|$ are separately divergent and only their difference is convergent [97].

It should be pointed out that in the typical case of experimental data with the squared modulus of amplitude reflectivity, i.e. reflectance, the conventional phase retrieval by K-K relation (4.35) may not determine the phase completely. This may happen, for instance, when recording reflection spectra through a transparent window [98], and also in the case of oblique TM-polarized incident light, provided that the system fulfils some constraints related to the dielectric function of the absorbing medium. Here we analyze the case of oblique incidence. In order to clarify the differences in phase retrieval from the logarithm of reflectance related to TM- and TE-polarized light, we next employ expressions (3.21), (3.24), and (3.28), and for the sake of simplicity, we assume that light is incident from air to the light absorbing medium. Then we can write the following expressions:

$$R_{\mathrm{TE}} = \left| \frac{\cos\varphi_{\mathrm{i}} - \sqrt{N^2 - \sin^2\varphi_{\mathrm{i}}}}{\cos\varphi_{\mathrm{i}} + \sqrt{N^2 - \sin^2\varphi_{\mathrm{i}}}} \right|^2 , \tag{4.38}$$

$$R_{\mathrm{TM}} = \left| \frac{N^2\cos\varphi_{\mathrm{i}} - \sqrt{N^2 - \sin^2\varphi_{\mathrm{i}}}}{N^2\cos\varphi_{\mathrm{i}} + \sqrt{N^2 - \sin^2\varphi_{\mathrm{i}}}} \right|^2 . \tag{4.39}$$

The idea is to study the complex zeros of reflectance in the upper half plane. From (4.38), we observe that if we set $R_{\mathrm{TE}} = 0$, this equation holds only if $N = 1$, and this same condition is also true for the case of the normal incidence. Thus, taking the natural logarithm, i.e. $\ln R_{\mathrm{TE}}$, is not a problem and the normal procedure of phase retrieval by (4.35) for both normal light incidence (polarization of the light is immaterial) and for oblique incidence, TE-polarized light is valid. On the contrary, the condition $R_{\mathrm{TM}} = 0$, which in the case of a nonabsorbing medium is achieved at Brewster angle φ_{i} such that $\tan\varphi_{\mathrm{i}} = \eta$, and is achieved also for a purely imaginary frequency. The reason is that the dielectric function $\varepsilon(\omega)$ of the medium is a monotonically decreasing real function of the frequency [38]. This means that $N^2 = \varepsilon \in \mathbb{R}$. Now allowing a purely real refractive index, which is taken along the imaginary frequency axis, yield according to (4.39) that there are complex zeros on the imaginary axis if

$$\varepsilon_{\mathrm{static}} \geq \tan^2\varphi_{\mathrm{i}} \geq \varepsilon_{\infty} , \tag{4.40}$$

where the upper limit is the static value of the dielectric function. Under condition (4.40), complex zeros of the reflectance of TM-polarized light yield singularities of the logarithm of reflectance. Such singular points represent branch points of the logarithmic function. Nevertheless, the phase in such a case can be obtained using the Blaschke product as devised by Toll [1,99]. If ω_j denotes complex zeros with $j = 1, \ldots, J$, then the Blaschke product is

$$B(\omega) = \prod_{j=1}^{J} \frac{\omega - \omega_j}{\omega - \omega_j^*} . \tag{4.41}$$

True reflectivity is obtained by multiplying the complex reflectivity, obtained by conventional K-K phase retrieval procedure, by the Blaschke product, as follows:

$$r_{\text{true}}(\omega) = r_{\text{KK}}(\omega) \prod_{j=1}^{J} \frac{\omega - \omega_j}{\omega - \omega_j^*}. \tag{4.42}$$

Grosse and Offermann [100] have studied such a phase correction in the context of TM-polarized light incident on bulk and layered media and provided a method for phase reconstruction from experimental data. In their method, they estimate the complex zeros. Phase retrieval from the reflectance of isotropic and anisotropic media, based on utilization of the K-K relation related to the logarithm of reflectivity, and corresponding phase corrections were studied by Yamamoto and Ishida [101,102]. We remark already here that if maximum entropy method techniques (discussed in Chap. 10) are applied, it is possible to work with the reflectivity function instead of considering its logarithm. This means that many mathematical complications related to the functional behavior of the logarithmic function can be avoided.

Reflection spectroscopy has for a long time constituted a basic method along with K-K relations for resolving a complex refractive index from reflectance. Note that in the case of a phase retrieval procedure based on relations (4.35) and (4.30), it is possible to obtain the absolute value of the refractive index. The reflection measurement mode is especially reasonable when studying opaque media and liquids. Naturally ellipsometry provides information on the complex refractive index. However, due to the limitations of the broadness of the wavelength ranges that ellipsometers provide, the simple reflectometer may be more useful. One reason is the broad spectral range that is available with a reflectometer, which is also not a sophisticated device if compared to an ellipsometer. In Fig. 4.6 we show the results of phase retrieval obtained using (4.35) of orthorhombic sulfur measured with light polarized parallel and perpendicular, respectively, to the projection of the c-axis of the (111) plane.

The unavoidable problem of having measured data in a limited spectral range hinders the efficacy of the data inversion procedure performed with K-K relations. Extrapolations beyond the measured range may sometimes be hazardous. In order to overcome this problem, Modine et al. [104] presented a method of coupling reflectometric and ellipsometric data in order to improve the success of the K-K analysis. Their analysis is based on the truncated K-K relation

$$\theta_{\text{KK}}(\omega) = -\frac{\omega}{\pi} \mathrm{P} \int_0^{\omega_{\text{m}}} \frac{\ln R(\omega')}{\omega'^2 - \omega^2} \mathrm{d}\omega', \tag{4.43}$$

where ω_{m} is the upper limit of the reflectance data, and they assumed that the reflectance data can be accurately extrapolated to zero frequency. The remainder of the K-K relation for the unmeasured angular frequency region was obtained by the use of a geometric series expansion as follows:

$$\Delta\theta(\omega) = -\frac{\omega}{\pi} \mathrm{P} \int_{\omega_{\text{m}}}^{\infty} \frac{\ln R(\omega')}{\omega'^2 - \omega^2} \mathrm{d}\omega' = \sum_{j=0}^{\infty} A_{2j+1} \omega^{2j+1}. \tag{4.44}$$

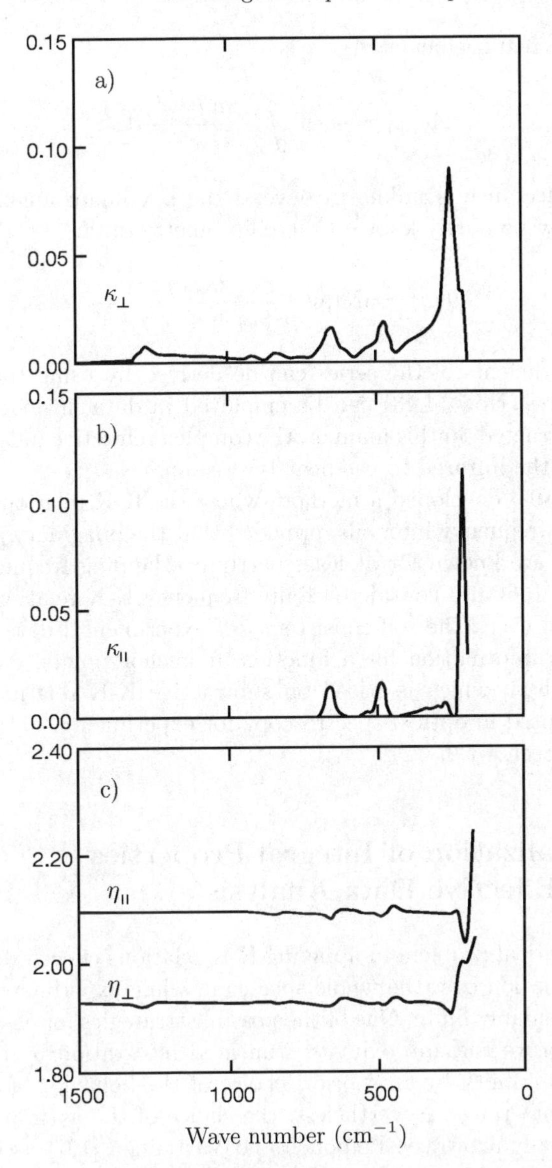

Fig. 4.6. (**a**) Perpendicular $\kappa_\perp$; (**b**) parallel extinction coefficient $\kappa_\parallel$, and (**c**) the refractive index η of orthorhombic sulfur retrieved from reflectance data. Reproduced from [103]

In (4.44), the real coefficient A_{2j+1} is

$$A_{2j+1} = -\frac{1}{\pi}\mathrm{P}\int_{\omega_{\mathrm{m}}}^{\infty} \frac{\ln R(\omega')}{\omega'^2}\mathrm{d}\omega'. \tag{4.45}$$

Although reflectance is unknown beyond the maximum angular frequency, the phase may be partly known from ellipsometry data:

$$\theta(\omega) = \arctan\frac{2\kappa(\omega)}{\eta^2(\omega) + \kappa^2(\omega) - 1}. \tag{4.46}$$

Then the coefficients of the series can de derived by using the the method of least squares. Now (4.43) can be employed in data inversion. Modine et al. [104] constructed, in this manner, the complex refractive index of tantalum carbide from the infrared to the near UV region.

Hulthén [105] developed a method, where the K-K relations are generalized for finite frequency intervals, provided that the imaginary and real parts of a function are known for at least partly overlapping frequency intervals. Milton et al. [106] also considered finite frequency K-K relations. They gave inequalities to check the self-consistency of experimental data, which, however, requires information for a function at anchor points. An alternative, powerful method, which is based on subtractive K-K relations and which has been adopted in optical spectroscopy for experimental data inversion, is presented in Sect. 4.7.2.

4.7 Generalization of Integral Properties for More Effective Data Analysis

A serious practical problem in applying K-K relations effectively is that they require information over the whole spectrum, while experimentally the data range is unavoidably finite. One of the possible strategies for easing this problem is to improve the approximate truncated integration by extending the integration to infinity by an a priori choice of the behavior of susceptibility outside the data range. Nevertheless, the choice of the asymptotic behavior cannot be totally arbitrary. Peiponen and Vartiainen [107] used a Gaussian line shape for the extinction coefficient in data extrapolation. They observed that calculated values of the real refractive index obtained from K-K analysis were erroneous. The errors were caused by the failure of the symmetry property of the Gaussian line shape, which is not an odd function.

In the next subsections, we present a more detailed description of two promising approaches that, by exploiting some refinements of conventional K-K relations, provide more efficient and practical tools for data analysis.

4.7.1 Generalized Kramers-Kronig Relations

In order to overcome the problem of the relatively low convergence of K-K relations, Altarelli and Smith [78] introduced generalized K-K relations by considering the functions $[N(\omega)-1]^m$ and $\omega^m[N(\omega)-1]^m$. All the functions belonging to both classes have the same holomorphic properties of $N(\omega)$, are square-integrable functions, and for a given m, the convergence is proportional to ω^{-2m} and ω^{-m}, respectively. The symmetry properties of the functions $\omega^m[N(\omega)-1]^m$ depend on the parity of the power ω^m. For odd numbers m, the generalized K-K relations can be written as follows:

$$\omega^m \mathrm{Re}\{[N(\omega)-1]^m\} = \frac{2\omega}{\pi}\mathrm{P}\int_0^\infty \frac{\omega'^m \mathrm{Im}\{[N(\omega')-1]^m\}}{\omega'^2-\omega^2}\mathrm{d}\omega', \tag{4.47}$$

$$\omega^m \mathrm{Im}\{[N(\omega)-1]^m\} = -\frac{2}{\pi}\mathrm{P}\int_0^\infty \frac{\omega'^{m+1} \mathrm{Re}\{[N(\omega')-1]^m\}}{\omega'^2-\omega^2}\mathrm{d}\omega'. \tag{4.48}$$

For even numbers m, the generalized K-K relations are of the form

$$\omega^m \mathrm{Re}\{[N(\omega)-1]^m\} = \frac{2}{\pi}\mathrm{P}\int_0^\infty \frac{\omega'^{m+1} \mathrm{Im}\{[N(\omega')-1]^m\}}{\omega'^2-\omega^2}\mathrm{d}\omega', \tag{4.49}$$

$$\omega^m \mathrm{Im}\{[N(\omega)-1]^m\} = -\frac{2\omega}{\pi}\mathrm{P}\int_0^\infty \frac{\omega'^m \mathrm{Re}\{[N(\omega')-1]^m\}}{\omega'^2-\omega^2}\mathrm{d}\omega'. \tag{4.50}$$

For values $m > 2$, the convergence of generalized K-K relations is strictly faster than that of conventional K-K relations, which is of crucial importance in experimental data analysis.

If we apply the superconvergence theorem to (4.47)–(4.50) and consider the asymptotic behavior (4.25) of $N(\omega)-1$, we derive a more general set of sum rules [78]:

$$\int_0^\infty \omega' \mathrm{Im}\{[N(\omega')-1]^m\}\mathrm{d}\omega' = \begin{cases} \frac{\pi}{4}\omega_\mathrm{p}^2 & m=1 \\ 0 & m>0 \end{cases}, \tag{4.51}$$

$$\int_0^\infty \omega'^{m-1} \mathrm{Re}\{[N(\omega')-1]^m\}\mathrm{d}\omega' = 0 \quad m=1,3,5,\cdots, \tag{4.52}$$

$$\int_0^\infty \omega'^m \mathrm{Im}\{[N(\omega')-1]^m\}\mathrm{d}\omega' = \begin{cases} \frac{\pi}{4}\omega_\mathrm{p}^2 & m=1 \\ 0 & m=3,5,7,\cdots \end{cases}, \tag{4.53}$$

$$\int_0^\infty \omega'^{m-2}\mathrm{Im}\{[N(\omega')-1]^m\}\mathrm{d}\omega' = 0 \quad m = 3,5,7,\cdots, \tag{4.54}$$

$$\int_0^\infty \omega'^{m}\mathrm{Re}\{[N(\omega')-1]^m\}\mathrm{d}\omega' = 0 \quad m = 2,4,6,\cdots, \tag{4.55}$$

$$\int_0^\infty \omega'^{m-2}\mathrm{Re}\{[N(\omega')-1]^m\}\mathrm{d}\omega' = 0 \quad m = 4,6,8,\cdots, \tag{4.56}$$

$$\int_0^\infty \omega'^{m-1}\mathrm{Im}\{[N(\omega')-1]^m\}\mathrm{d}\omega' = 0 \tag{4.57}$$

$$\int_0^\infty \omega'^{m+1}\mathrm{Im}\{[N(\omega')-1]^m\}\mathrm{d}\omega' = \begin{cases} -\frac{\pi}{8}\omega_\mathrm{p}^4 & m=2 \\ 0 & m=4,6,8,\cdots \end{cases}, \tag{4.58}$$

Similar results could also be obtained for higher powers of linear susceptibility. Peiponen and Asakura [72] presented the generalized K-K relations and sum rules for layered nanostructures by considering both even and odd powers of the function $\omega^m[\chi_{\mathrm{eff}}^{(1)}(\omega)]^m$. Their results, nevertheless, apply to all of the nanostructural composites considered here.

Along the lines of Altarelli and Smith [78], Smith and Manogue [95] considered the powers of complex reflectivity. They proved that functions $[r(\omega)]^m$ and $\omega^m[r(\omega)]^m$ are square-integrable functions with proper asymptotic behavior which allow the derivation of mixed K-K relations. These K-K relations mix both the amplitude and the phase of complex reflectivity. The real and imaginary parts of the function $[r(\omega)]^m$ result in the following:

$$|r(\omega)|^m\cos[m\theta(\omega)] = \frac{2}{\pi}\mathrm{P}\int_0^\infty \frac{\omega|r(\omega')|^m\sin[m\theta(\omega')]}{\omega'^2-\omega^2}\mathrm{d}\omega', \tag{4.59}$$

$$|r(\omega)|^m\sin[m\theta(\omega)] = -\frac{2\omega}{\pi}\mathrm{P}\int_0^\infty \frac{|r(\omega')|^m\cos[m\theta(\omega')]}{\omega'^2-\omega^2}\mathrm{d}\omega'. \tag{4.60}$$

In a similar manner, Smith and Manogue [95] derived generalized mixed relations for even and odd powers of $\omega^m[r(\omega)]^m$, which are analogous to (4.48)–(4.50). Mixed K-K relations can be used in measurements where both the amplitude and the phase of the reflected beam are obtained. Moreover, they can be used in the derivation of various sum rules that provide tools for testing experimental data such as those found in ellipsometric studies.

4.7.2 Subtractive K-K Relations

Bachrach and Brown [108] introduced singly subtractive K-K relations (SSKK) in order to reduce the errors caused by the finite spectrum as they calculated the real refractive index of a medium from a measured absorption spectrum. They measured one reference point (or anchor point) independently, which was used to improve the accuracy of K-K analysis. Conventional K-K relation (4.26) can be written at the reference point, say, at frequency ω_1, in the form

$$\eta(\omega_1) - 1 = \frac{2}{\pi}\mathrm{P}\int_0^\infty \frac{\omega'\kappa(\omega')}{\omega'^2 - \omega_1^2}\mathrm{d}\omega', \tag{4.61}$$

where $\eta(\omega_1)$ is known a priori. By subtracting (4.61) from (4.26), Bachrach and Brown [108] obtained a singly subtractive K-K relation as follows:

$$\begin{aligned} \eta(\omega) - \eta(\omega_1) &= \frac{2}{\pi}\mathrm{P}\int_0^\infty \frac{\omega'\kappa(\omega')}{\omega'^2 - \omega^2}\mathrm{d}\omega' - \frac{2}{\pi}\mathrm{P}\int_0^\infty \frac{\omega'\kappa(\omega')}{\omega'^2 - \omega_1^2}\mathrm{d}\omega' \\ &= \frac{2}{\pi}\left[\mathrm{P}\int_0^\infty \frac{[\omega'^2 - \omega_1^2]\omega'\kappa(\omega') - [\omega^2 - \omega^2]\omega'\kappa(\omega')}{(\omega'^2 - \omega^2)(\omega'^2 - \omega_1^2)}\mathrm{d}\omega'\right] \\ &= \frac{2}{\pi}\mathrm{P}\int_0^\infty \frac{[\omega^2 - \omega_1^2]\omega'\kappa(\omega')}{(\omega'^2 - \omega^2)(\omega'^2 - \omega_1^2)}\mathrm{d}\omega' \\ &= \frac{2(\omega^2 - \omega_1^2)}{\pi}\mathrm{P}\int_0^\infty \frac{\omega'\kappa(\omega')}{(\omega'^2 - \omega^2)(\omega'^2 - \omega_1^2)}\mathrm{d}\omega'. \end{aligned} \tag{4.62}$$

The subtractive form of a K-K integral converges more rapidly than a conventional K-K relation. Singly subtractive K-K relations have been used in improving the convergence of the phase of a measured reflectancespectrum [109]. Recently, Palmer et al. [110] described multiply subtractive K-K relations (MSKK) in order to obtain the optical constants of a medium with a single reflectance measurement with a finite frequency range. Multiply subtractive K-K relations are derived in a manner similar to singly subtractive K-K relations. MSKK analysis requires several reference points in the phase retrieval procedure. For reflection spectroscopy, the reference points are the discrete values of the complex refractive index, which are obtained from independent measurement and belong to the measured frequency range. A doubly subtractive K-K relation is derived with the aid of (4.62), by replacing the first reference point ω_1 with the second reference point ω_2, which becomes

$$\eta(\omega) - \eta(\omega_2) = \frac{2(\omega^2 - \omega_2^2)}{\pi}\mathrm{P}\int_0^\infty \frac{\omega'\kappa(\omega')}{(\omega'^2 - \omega^2)(\omega'^2 - \omega_2^2)}\mathrm{d}\omega'. \tag{4.63}$$

Equations (4.62) and (4.63) are subtracted, and the result is analogous to the results of Palmer et al. [110] for the phase of a measured amplitude. The doubly subtracted K-K relation for the real refractive index is of the form

$$\frac{\eta(\omega)-1}{(\omega^2-\omega_1^2)(\omega^2-\omega_2^2)}-\frac{\eta(\omega_1)-1}{(\omega^2-\omega_1^2)(\omega_1^2-\omega_2^2)}-\frac{\eta(\omega_2)-1}{(\omega^2-\omega_2^2)(\omega_1^2-\omega_2^2)}$$
$$=\frac{2}{\pi}\mathrm{P}\int_0^\infty\frac{\omega'\kappa(\omega')}{(\omega'^2-\omega^2)(\omega'^2-\omega_1^2)(\omega'^2-\omega_2^2)}\mathrm{d}\omega'. \tag{4.64}$$

In a similar manner, singly and doubly subtracted K-K relations can be generalized for arbitrary times subtracted K-K relations. The general formula for Q times subtracted K-K relations for the reflectancespectrum was presented in the literature [110]. Here similar formulas are presented for the refractive index:

$$\eta(\omega)-1=\left[\frac{(\omega^2-\omega_2^2)(\omega^2-\omega_3^2)\cdots(\omega^2-\omega_Q^2)}{(\omega_1^2-\omega_2^2)(\omega_1^2-\omega_3^2)\cdots(\omega_1^2-\omega_Q^2)}\right][\eta(\omega_1)-1]+\cdots$$
$$+\left[\frac{(\omega^2-\omega_1^2)\cdots(\omega^2-\omega_{j-1}^2)(\omega^2-\omega_{j+1}^2)\cdots(\omega^2-\omega_Q^2)}{(\omega_j^2-\omega_1^2)\cdots(\omega_j^2-\omega_{j-1}^2)(\omega_j^2-\omega_{j+1}^2)\cdots(\omega_j^2-\omega_Q^2)}\right][\eta(\omega_j)-1]+\cdots$$
$$+\left[\frac{(\omega^2-\omega_1^2)(\omega^2-\omega_2^2)\cdots(\omega^2-\omega_{Q-1}^2)}{(\omega_Q^2-\omega_1^2)(\omega_Q^2-\omega_2^2)\cdots(\omega_Q^2-\omega_{Q-1}^2)}\right][\eta(\omega_Q)-1]$$
$$+\frac{2}{\pi}\left[(\omega^2-\omega_1^2)(\omega^2-\omega_2^2)\cdots(\omega^2-\omega_Q^2)\right]\mathrm{P}\int_0^\infty\frac{\omega'\kappa(\omega')\mathrm{d}\omega'}{(\omega'^2-\omega^2)\cdots(\omega'^2-\omega_Q^2)}, \tag{4.65}$$

$$\kappa(\omega)=\left[\frac{(\omega^2-\omega_2^2)(\omega^2-\omega_3^2)\cdots(\omega^2-\omega_Q^2)}{(\omega_1^2-\omega_2^2)(\omega_1^2-\omega_3^2)\cdots(\omega_1^2-\omega_Q^2)}\right][\eta(\omega_1)-1]+\cdots$$
$$+\left[\frac{(\omega^2-\omega_1^2)\cdots(\omega^2-\omega_{j-1}^2)(\omega^2-\omega_{j+1}^2)\cdots(\omega^2-\omega_Q^2)}{(\omega_j^2-\omega_1^2)\cdots(\omega_j^2-\omega_{j-1}^2)(\omega_j^2-\omega_{j+1}^2)\cdots(\omega_j^2-\omega_Q^2)}\right][\eta(\omega_j)-1]+\cdots$$
$$+\left[\frac{(\omega^2-\omega_1^2)(\omega^2-\omega_2^2)\cdots(\omega^2-\omega_{Q-1}^2)}{(\omega_Q^2-\omega_1^2)(\omega_Q^2-\omega_2^2)\cdots(\omega_Q^2-\omega_{Q-1}^2)}\right][\eta(\omega_Q)-1]$$
$$+\frac{2}{\pi}\left[(\omega^2-\omega_1^2)(\omega^2-\omega_2^2)\cdots(\omega^2-\omega_Q^2)\right]\mathrm{P}\int_0^\infty\frac{[\eta(\omega)-1]\mathrm{d}\omega'}{(\omega'^2-\omega^2)\cdots(\omega'^2-\omega_Q^2)}. \tag{4.66}$$

Similar results can also be obtained for linear susceptibility as well as for the logarithm of linear reflectivity. Multiply subtractive K-K relations have very rapid convergence, which significantly reduces the errors caused by extrapolation. With a suitable choice of reference points, the optical constants of the medium can be determined by a single reflectance measurement [110].

5 General Properties of the Nonlinear Optical Response

5.1 Nonlinear Optics: A Brief Introduction

Linear optics provides a complete description of light–matter interaction only in the limit of weak radiation sources. When we consider more powerful radiation sources, the phenomenology of light–matter interaction is much more complex and interesting. Entirely new classes of processes, related to the effects of nonlinearity in the interaction, can be observed experimentally. Therefore, new theoretical tools have to be introduced in order to account for these phenomena.

When matter interacts with one or more intense monochromatic light sources, a very notable nonlinear effect is *frequency mixing* i.e. the presence in the spectrum of the outgoing radiation of nonvanishing frequency components corresponding to the sum and difference of the frequency components of the incoming radiation. This process is greatly enhanced when either the incoming radiation or their combinations are in resonance with the permitted transitions of the material. Within the large family of frequency mixing phenomena, a very relevant process is *harmonic on generation.* Harmonic generation is the production of outgoing radiation with relevant spectral components at frequencies that are integral multiples of the frequencies of the incoming monochromatic light sources. Another very relevant family of nonlinear effects is *multi–photon absorption* phenomena, i.e. processes where the electronic transitions are assisted by the simultaneous intervention of more than one photon. In recent years, nonlinear optical investigations have gained fundamental importance in the study of the properties of gases, liquids, and condensed matter, and they have assumed great relevance in technological applications as well as in the study of organic and biological materials.

In 1931, Goppert–Mayer [41] proposed that two-photon assisted transitions between two states were theoretically possible, but only in the 1950s, when suitable monochromatic radiation sources became available in the regions of radiowaves and microwaves, were nonlinear processes observed. Hughes et al. [111] measured the two-photon transition between two hyperfine levels, and Battaglia et al. [112] observed the two-photon transition between the rotational levels of the carbon oxy-sulphide (OCS) molecule.

In the 1960s, the advent of laser technology permitted observation of nonlinear phenomena at optical frequencies. In 1961, Kaiser and Garrett [113] observed the first two-photon transitions in solid matter and in the same year, Franken et al. [114] experimentally proved the possibility of second-harmonic on generation. These experiments had a great impact on the scientific community and provided a strong impulse for theoretical and experimental research in nonlinear optics. Woodbury and Ng [115] observed the phenomenon of stimulated Raman absorption [116, 117], Marker et al. [118] reported on third-harmonic-generation, and Singh and Bradley [119] observed the three-photon absorption process. A book by Bloembergen [17] provides a complete report of the scientific activity in this rapidly growing branch of physics, and recently a complete collection of the relevant papers from the early days of nonlinear optics was published [120].

Exploiting the phenomena of frequency mixing in suitable materials and taking advantage of the transitional resonances in order to obtain good power conversion, it was possible to obtain new coherent and monochromatic sources with peak frequencies different from those of the primary sources, and thus increase the possibilities of the experimental study of nonlinear optical phenomena. In the 1970s, tunable monochromatic light sources became available thanks to the development of solid-state and dye lasers [121, 122]. This permitted nonlinear spectroscopic studies that could span a certain frequency range, instead of being limited to given discrete frequencies. The passage from discrete to a continuum greatly improved the investigation of the nonlinear optical properties of materials.

Thereafter, the development of the mode-locked laser permitted the generation of pulses shorter than 10^{-9}s with very high peak power. This permitted experimental study of transient nonlinear phenomena and of nonlinear effects of a higher order. Since the 1980s, lasers with peak intensities of 10^{13} W/cm^2 or more have become available, so that very high order nonlinear processes, such as the production of harmonics of an order higher than 100, have been possible [123–125].

The theory of nonlinear optics is based mostly on a perturbative approach introduced by Bloembergen [17], which is essentially a semiclassical analogue of the Feynman graphs formalism [16] and which generally adopts dipolar approximation. In the perturbative approach, the nonlinear optical properties of a material are fully described by nonlinear susceptibilities, which are obtained by applying the Fourier transform to nonlinear Green functions, which describe the higher order dynamics of the system. The nonlinear Green function can be obtained using a generalization of Kubo's linear response theory [126]. Results obtained using the perturbative theory are generally in good agreement with the experimental data unless we consider the interaction of matter with ultra-intense lasers, such as those responsible for very high order harmonic-generation. A detailed presentation of the perturbative theory for nonlinear optics and of comparison with experimental data can be found in recent books by Butcher and Cotter [18] and Boyd [19].

In this chapter, we will review the main points of the material presented in the past decade dealing with the establishment of general integral properties for nonlinear optical susceptibilities. We extend the results obtained in Chap. 3 to higher orders of perturbations. We follow the perturbative approach to describe nonlinear optical processes, and we adopt the general formalism of the density matrix [11]. We also present a discussion of nonlinear local field effects and of how they can be described with the effective medium approximation for both homogeneous and nanostructured media.

5.2 Nonlinear Optical Properties

In our analysis, we assume the dipolar approximation and generalize the definition of polarization presented in Chap. 3 to all orders of nonlinearity. Following (2.34), total polarization can be expressed as the sum of linear and nonlinear contributions:

$$\boldsymbol{P}(t) = \boldsymbol{P}^{(1)}(t) + \boldsymbol{P}^{\mathrm{NL}}(t), \tag{5.1}$$

where total nonlinear polarization $\boldsymbol{P}^{\mathrm{NL}}(t)$ is defined as the sum over the various orders of nonlinearity:

$$\boldsymbol{P}^{\mathrm{NL}}(t) = \sum_{n=2}^{\infty} \boldsymbol{P}^{(n)}(t). \tag{5.2}$$

We define nth-order nonlinear polarization as a multiple convolution:

$$P_i^{(n)}(t) = \int_{-\infty}^{\infty} G_{ij_1 \ldots j_n}^{(n)}(t_1 \ldots t_n) E_{j_1}(t - t_1) \ldots E_{j_n}(t - t_n) \mathrm{d}t_1 \ldots \mathrm{d}t_n, \tag{5.3}$$

where the nonlinear Green function is symmetrical in time variables and respects the temporal causality principle

$$\begin{aligned} &G_{ij_1 \ldots j_i \ldots j_k \ldots j_n}^{(n)}(t_1, \ldots, t_i, \ldots, t_k, \ldots, t_n) \\ &\qquad = G_{ij_1 \ldots j_k \ldots j_i \ldots j_n}^{(n)}(t_1, \ldots, t_k, \ldots, t_i, \ldots, t_n), \end{aligned} \tag{5.4}$$

$$G_{ij_1 \ldots j_n}(t_1, \ldots, t_n) = 0, \quad t_i < 0, \quad 1 \le i \le n. \tag{5.5}$$

Computing the Fourier transform of expression (5.3), we obtain

$$\begin{aligned} P_i^{(n)}(\omega) = &\int_{-\infty}^{\infty} \chi_{ij_1 \ldots j_n}^{(n)}\left(\sum_{l=1}^{n} \omega_l; \omega_1, \ldots, \omega_n\right) E_{j_1}(\omega_1) \ldots E_{j_n}(\omega_n) \\ &\times \delta\left(\omega - \sum_{l=1}^{n} \omega_l\right) \mathrm{d}\omega_1 \ldots \mathrm{d}\omega_n, \end{aligned} \tag{5.6}$$

where the Dirac delta function δ guarantees that the sum of the arguments of the Fourier transforms of the electric field equals the argument of the Fourier transform of polarization. This means that each ω component of the nth-order nonlinear polarization results from the interaction of n photons mediated by the material under examination, which is described by the nonlinear susceptibility function, defined as

$$\begin{aligned}&\chi^{(n)}_{ij_1\cdots j_n}\left(\sum_{l=1}^{n}\omega_l;\omega_1,\ldots,\omega_n\right)\\&\quad=\int_{-\infty}^{\infty}G^{(n)}_{ij_1\cdots j_n}(t_1,\ldots,t_n)\exp\left[\mathrm{i}\sum_{l=1}^{n}\omega_l t_l\right]\mathrm{d}t_1\ldots\mathrm{d}t_n.\end{aligned}\tag{5.7}$$

It is customary in the literature [18, 19] to set as the first argument of the susceptibility the sum of all the frequency arguments before the semicolon. From definition (5.7) it is possible to derive the a priori properties of the nonlinear susceptibility functions by considering the properties of the nonlinear Green function described in expressions (5.4). The symmetry of the nonlinear Green function with respect to the time variable exchange implies that nonlinear susceptibility functions obey a symmetrical property for the frequency variables

$$\begin{aligned}&\chi^{(n)}_{ij_1\cdots j_i\cdots j_k\cdots j_n}\left(\sum_{l=1}^{n}\omega_l;\omega_1,\ldots,\omega_i,\ldots\omega_k,\ldots,\omega_n\right)\\&\quad=\chi^{(n)}_{ij_1\cdots j_k\cdots j_i\cdots j_n}\left(\sum_{l=1}^{n}\omega_l;\omega_1,\ldots,\omega_k,\ldots\omega_i,\ldots,\omega_n\right)\end{aligned}\tag{5.8}$$

for all $i,j\in\{1,\ldots,n\}$. Since $\boldsymbol{P}^{(n)}(t)$ and the nonlinear Green function are real, we derive, along the lines of the previously analyzed linear case, that at every order n, the following relation holds:

$$\chi^{(n)}_{ij_1\cdots j_n}\left(\sum_{l=1}^{n}-\omega_l;-\omega_1,\ldots,-\omega_n\right)=\left[\chi^{(n)}_{ij_1\cdots j_n}\left(\sum_{l=1}^{n}\omega_l;\omega_1,\ldots,\omega_n\right)\right]^*,\tag{5.9}$$

so that, separating the real and imaginary parts, we can write

$$\begin{aligned}&\mathrm{Re}\left\{\chi^{(n)}_{ij_1\cdots j_n}\left(\sum_{l=1}^{n}-\omega_l;-\omega_1,\ldots,-\omega_n\right)\right\}\\&\quad=\mathrm{Re}\left\{\chi^{(n)}_{ij_1\cdots j_n}\left(\sum_{l=1}^{n}\omega_l;\omega_1,\ldots,\omega_n\right)\right\},\end{aligned}\tag{5.10}$$

$$
\begin{aligned}
-\operatorname{Im}\left\{\chi^{(n)}_{ij_1\ldots j_n}\left(\sum_{l=1}^{n}-\omega_l;-\omega_1,\ldots,-\omega_n\right)\right\}\\
=\operatorname{Im}\left\{\chi^{(n)}_{ij_1\ldots j_n}\left(\sum_{l=1}^{n}\omega_l;\omega_1,\ldots,\omega_n\right)\right\}.
\end{aligned}
\tag{5.11}
$$

However, we wish to emphasize that it has been suggested that in nonlinear optics, the nonlinear Green function may also take a complex form [127,128].

We consider the realistic case of the interaction of matter with m monochromatic plane waves with different amplitude, polarization, and frequency:

$$
E_j(t)=\sum_{i=1}^{m}E_j^{(i)}\exp[-\mathrm{i}\omega_i t]+c.c.,
\tag{5.12}
$$

where $E_j^{(i)}$ is the amplitude of the electric field of the ith wave in the j-direction, so that:

$$
E_j(\omega)=\sum_{i=1}^{m}2\pi E_j^{(i)}\delta(\omega-\omega_i)+2\pi E_j^{(i)}\delta(\omega+\omega_i).
\tag{5.13}
$$

From formulas (5.4) and (5.13), we deduce that each spectral component of the nth-order polarization is given by the sum ω_Σ of n among the $2m$ (m positive and m negative) frequencies characterizing the spectrum of incoming radiation:

$$
\begin{aligned}
P_i^{(n)}(\omega)=\sum_{\Omega_i\in\{\pm\omega_1,\ldots\pm\omega_m\}}\chi^{(n)}_{ij_1\ldots j_n}\left(\sum_{l=1}^{n}\Omega_l;\Omega_1,\ldots,\Omega_n\right)\\
\times E_{j_1}^{(\Omega_1)}\ldots E_{j_n}^{(\Omega_n)}2\pi\delta\left(\omega-\sum_{i=1}^{n}\Omega_i\right),
\end{aligned}
\tag{5.14}
$$

where $E_{j_i}^{(\Omega_i)}=E_{j_i}^{(i)}$ if $\Omega_i=\pm\omega_i$. Separating each frequency component in expression (5.14), we obtain

$$
P_i^{(n)}(\omega)=\sum_{\{\omega_\Sigma\}}\overline{P}_i^{(n)}(\omega_\Sigma)\,2\pi\delta(\omega-\omega_\Sigma),
\tag{5.15}
$$

where we are summing over all possible distinct values $\{\omega_\Sigma\}$ of the sum of n among the $2m$ frequencies in the electric field spectrum. The coefficients $\overline{P}_i^{(n)}(\omega_\Sigma)$ are such that

$$
P_i^{(n)}(t)=\sum_{\omega_\Sigma}\overline{P}_i^{(n)}(\omega_\Sigma)\exp[-\mathrm{i}\omega_\Sigma t].
\tag{5.16}
$$

The presence in nonlinear polarization of all possible combinations frequencies of the incoming radiation is commonly referred to as frequency mixing.

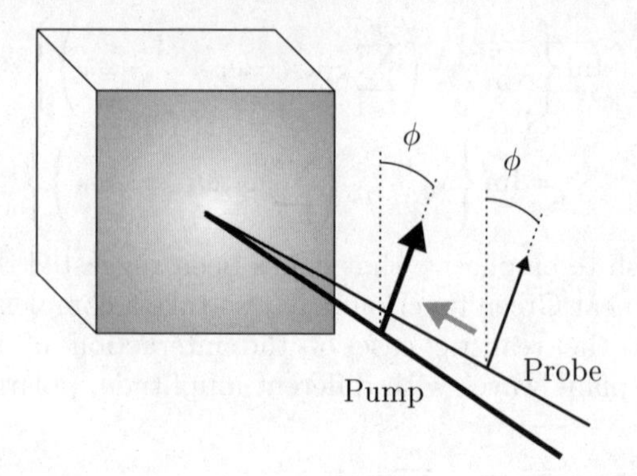

Fig. 5.1. Scheme of a pump-and-probe experiment where the two laser beams have the same polarization unit vectors. *Black line*: pump laser; *dotted line*: probe laser

5.2.1 Pump-and-Probe Processes

Two highly relevant nonlinear physical phenomena that can be described using this quite general formalism are harmonic-generation processes and pump-and-probe processes. We will deal extensively with harmonic-generation processes in the next chapters. Here we consider the relevant and instructive example of pump-and-probe processes. We consider the related though experimental setting, which is depicted in Fig. 5.1, where two lasers are present. We make the simplifying assumption that both light beams are linearly polarized in the x-direction:

$$\boldsymbol{E}(t) = \hat{x}E^{(1)} \exp[-\mathrm{i}\omega_1 t] + \hat{x}E^{(2)} \exp[-\mathrm{i}\omega_2 t] + c.c. \tag{5.17}$$

We assume that index (1) refers to the probe laser, whose frequency can be changed, and index (2) to the pump laser, whose frequency is fixed. Then,

$$E^{(2)} \gg E^{(1)}. \tag{5.18}$$

We wish to consider the expression of the polarization of the third order at frequency $\omega_\Sigma = \omega_1$, i.e. the first variation in the linear optical response at the frequency of the probe laser due to light–matter coupling. Specifying $k = 3$ and $\omega = \omega_1$ in expression (5.15) and considering the form (5.17) of the incoming radiation, we obtain

$$\begin{aligned} \overline{P}^{(3)}(\omega_1) = {} & 6E^{(1)}\left(E^{(2)}\right)^2 \chi^{(3)}_{ixxx}(\omega_1;\omega_1,\omega_2,-\omega_2) \\ & + 3\left(E^{(1)}\right)^3 \chi^{(3)}_{ixxx}(\omega_1;\omega_1,\omega_1,-\omega_1), \end{aligned} \tag{5.19}$$

where the numeric coefficients can be deduced using combinatorial arguments, considering that the symmetry relation (5.8) holds. We underline that since the probe laser is much less intense than the pump laser, the first term is largely dominant in expression (5.19).

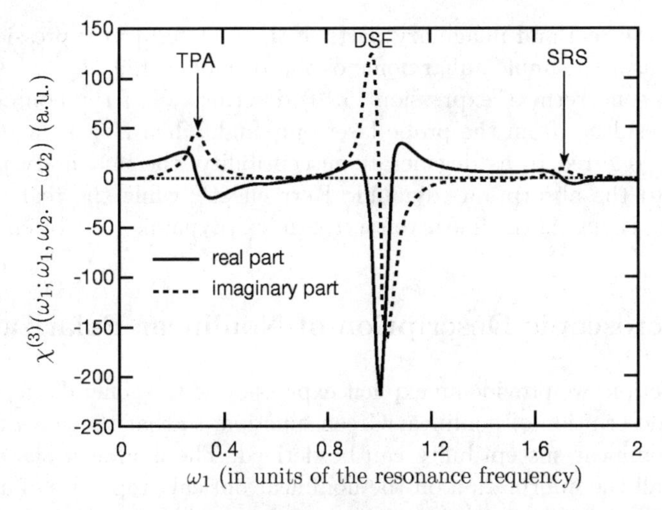

Fig. 5.2. Real and imaginary parts of the pump-and-probe susceptibility from a simplified model. The peaks corresponding to the dynamic Stark effect (DSE), two-photons absorption (TPA), and stimulated Raman scattering (SRS) are indicated. Reproduced from [129]

We emphasize that, since expression (5.19) represents the first nonlinear correction of the linear response at the probe laser frequency, it is possible to derive a correspondence between the nonlinear change in the index of refraction and the pump-and-probe susceptibility:

$$\begin{aligned} N^{\mathrm{NL}}(\omega_1,\omega_2,E_1,E_2) &= \Delta N(\omega_1) = \Delta\eta(\omega_1) + \mathrm{i}\Delta\kappa(\omega_1) \\ &= 4\pi\frac{6\left[E^{(2)}\right]^2\chi^{(3)}\left(\omega_1;\omega_1,\omega_2,-\omega_2\right) + 3\left[E^{(1)}\right]^2\chi^{(3)}\left(\omega_1;\omega_1,\omega_1,-\omega_1\right)}{2N(\omega_1)} \\ &\sim 4\pi\frac{3\left[E^{(2)}\right]^2\chi^{(3)}\left(\omega_1;\omega_1,\omega_2,-\omega_2\right)}{N(\omega_1)} = N^{\mathrm{NL}}(\omega_1,\omega_2,E_2), \end{aligned} \tag{5.20}$$

where we have used a scalar notation and we have simplified the expression by using condition (5.18). We underline that in expression (5.20), $N(\omega_1)$ is the linear index of refraction introduced in expression (3.16), and we have used the Δ symbol to emphasize that $N^{\mathrm{NL}}(\omega_1,\omega_2,E_2)$ is the first correction of the linear refraction index.

The dominant term of the quantity (5.19) – and correspondingly, of the quantity (5.20) – is especially relevant in physical terms since the imaginary part is related to the first-order nonlinear induced absorption, which includes the dissipative phenomena of two-photon absorption [42, 43, 130–132], stimulated Raman scattering [133–136], and dynamic Stark effect [137–141], while the real part describes nonlinear dispersive effects [14]. Figure 5.2 shows the values of the real and imaginary parts of the first term in expression (5.19) obtained using a simple anharmonic oscillator model [129].

The second term of expression (5.19) describes all of the nonlinear phenomena resulting from the probe laser only and, when $E^{(1)}$ is factored out, is usually referred to as degenerate susceptibility: the imaginary part contributes to the absorption (dynamic Kerr effect), while the real part contributes to changing the linear refractive index (dynamic Kerr birefringence).

5.3 Microscopic Description of Nonlinear Polarization

In this section, we provide an explicit expression of the general quantum mechanical description of nonlinear Green function, so that the general expression of nonlinear susceptibility can be derived. The nonlinear susceptibility contains all the information on the nonlinear optical properties of a system.

Following (2.34), we define the nth-order nonlinear polarization as the expectation value [17–19] of the electric dipole moment per unit volume over the nth-order perturbative term of the density operator (2.32):

$$P_i^{(n)}(t) \equiv \frac{1}{V}\mathrm{Tr}\left\{\sum_{\alpha=1}^{N} -e r_i^{\alpha} \rho^{(n)}(t)\right\}. \tag{5.21}$$

Therefore, we consider the nth differential equation in the concatenated system (2.33):

$$\mathrm{i}\hbar\partial_t \rho^{(n)}(t) = \left[H_0, \rho^{(n)}(t)\right] + \left[H_I^t, \rho^{(n-1)}(t)\right], \tag{5.22}$$

with the boundary condition $\rho^{(n)}(0) = 0$. The solution of this equation can be obtained recursively by considering each increasing perturbation order [18, 19], so that $\rho^{(n)}(t)$ can be expressed only as a function of the dipole operators and of $\rho(0)$:

$$\begin{aligned}\rho^{(n)}(t) = &\left(\frac{e}{-\mathrm{i}\hbar}\right)^n \int_{-\infty}^{t} \dots \int_{-\infty}^{t_{n-1}} E_{j_1}(t_1)\dots E_{j_n}(t_n) \\ &\times \left[\sum_{\alpha=1}^{N} r_{j_1}^{\alpha}(t_1 - t), \dots \left[\sum_{\alpha=1}^{N} r_{j_n}^{\alpha}(t_n - t), \rho(0)\right]\dots\right] \mathrm{d}t_1 \dots \mathrm{d}t_n .\end{aligned} \tag{5.23}$$

Considering that the following relation holds [18]:

$$\mathrm{Tr}\left\{\sum_{\alpha=1}^{N} r_i^\alpha \left[\sum_{\alpha=1}^{N} r_{j_1}^\alpha(-t_1), \ldots \left[\sum_{\alpha=1}^{N} r_{j_n}^\alpha(-t_n), \rho(0)\right] \ldots\right]\right\}$$
$$= (-1)^n \,\mathrm{Tr}\left\{\left[\sum_{\alpha=1}^{N} r_{j_n}^\alpha(-t_n), \ldots, \left[\sum_{\alpha=1}^{N} r_{j_1}^\alpha(-t_1), \sum_{\alpha=1}^{N} r_i^\alpha\right] \ldots\right] \rho(0)\right\}, \tag{5.24}$$

we can express the nth-order nonlinear polarization as the convolution of a nonlinear Green function multiplied by n electric fields:

$$P_i^{(n)}(t) = \int_{-\infty}^{\infty} \ldots \int_{-\infty}^{\infty} G_{ij_1 \ldots j_n}^{(n)}(t_1, \ldots, t_n) \times E_{j_1}(t-t_1) \ldots E_{j_n}(t-t_n)\, \mathrm{d}t_1 \ldots \mathrm{d}t_n, \tag{5.25}$$

where the nth-order nonlinear Green function is

$$G_{ij_1 \ldots j_n}^{(n)}(t_1, \ldots, t_n) = -\frac{e^{n+1}}{V(-\mathrm{i}\hbar)^n} \theta(t_1) \ldots \theta(t_n - t_{n-1})$$
$$\times \mathrm{Tr}\left\{\left[\sum_{\alpha=1}^{N} r_{j_n}^\alpha(-t_n), \ldots, \left[\sum_{\alpha=1}^{N} r_{j_i}^\alpha(-t_1), \sum_{\alpha=1}^{N} r_i^\alpha\right] \ldots\right] \rho(0)\right\}. \tag{5.26}$$

Therefore, the nth-order nonlinear polarization $\boldsymbol{P}^{(n)}(t)$ can generally be expressed as the expectation value of a t-dependent function over the thermodynamic equilibrium unperturbed density operator of the physical system under analysis, as obtained in the previous chapter for the linear case. All optical properties – linear and nonlinear – of a given physical system then depend only on its statistically relevant ground state.

Using definition (5.7), the nth-order nonlinear susceptibility of a general quantum system can then be expressed as follows:

$$\chi_{ij_1 \ldots j_n}^{(n)}\left(\sum_{l=1}^{n} \omega_j; \omega_1, \ldots, \omega_n\right)$$
$$= \int_{-\infty}^{\infty} G_{ij_1 \ldots j_n}^{(n)}(t_1, \ldots, t_n) \exp\left[\mathrm{i}\sum_{l=1}^{n} \omega_l t_l\right] \mathrm{d}t_1 \ldots \mathrm{d}t_n$$
$$= -\frac{e^{n+1}}{V(-\mathrm{i}\hbar)^n} \int_{-\infty}^{\infty} \theta(\tau_1) \ldots \theta(t_n - t_{n-1}) \exp\left[\mathrm{i}\sum_{l=1}^{n} \omega_l t_l\right]$$
$$\times \mathrm{Tr}\left\{\left[\sum_{\alpha=1}^{N} r_{j_n}^\alpha(-t_n), \ldots, \left[\sum_{\alpha=1}^{N} r_{j_1}^\alpha(-t_1), \sum_{\alpha=1}^{N} r_i^\alpha\right] \ldots\right] \rho(0)\right\} \mathrm{d}t_1 \ldots \mathrm{d}t_n. \tag{5.27}$$

In the linear case, the study of asymptotic behavior reveals that only the zeroth moment of the susceptibility has asymptotic behavior fast enough for the dispersion relation to converge. Hence, only one pair of independent dispersion relations – and of sum rules – can be established. It is reasonable to guess that the asymptotic decrease in nonlinear susceptibility for large values of each frequency variable is strictly faster than for linear susceptibility. The conceptual reason is that the electrons interacting with very energetic photons behave indistinguishably from the components of a free electron gas, since on the very short timescale relevant to the light–matter interaction, the other forces affecting the electrons are negligible. This is in conjunction with the observation that free electron gas behavior is already asymptotically saturated by considering the linear susceptibility alone, as seen in the Chap. 3.

5.4 Local Field and Effective Medium Approximation in Nonlinear Optics

In this section, we extend the analysis performed in Sect. 3.4 by analyzing how the local field effects influence the nonlinear optical properties of materials. We first present a general theory covering the case of homogeneous media and then focus our attention on nanostructured media. We show how the consideration of local field effects can be used to propose new nanostructured materials having well-defined optical properties.

5.4.1 Homogeneous Media

The nonlinear optical response of the elementary component of a physical system is thoroughly described by suitable microscopic hyperpolarizability $\beta^{(n)}$ [18,19], which connects the nonlinear polarization $p^{(n)}$ of the element to the local electric field acting on it:

$$p^{(n)}\left(\sum_{j=1}^{n}\omega_j\right) = \beta^{(n)}\left(\sum_{j=1}^{n}\omega_j;\omega_1,\ldots,\omega_n\right) E^{\mathrm{loc}}(\omega_1)\cdot E^{\mathrm{loc}}(\omega_n). \tag{5.28}$$

Along the line of the linear case, the macroscopic nonlinear polarization results are

$$\begin{aligned}\boldsymbol{P}^{(n)}\left(\sum_{j=1}^{n}\omega_j\right) &= \frac{1}{V}\int_V p^{(n)}\left(\sum_{j=1}^{n}\omega_j\right)\mathrm{d}v \\ &= \aleph\beta^{(n)}\left(\sum_{j=1}^{n}\omega_j;\omega_1,\ldots,\omega_n\right) E^{\mathrm{loc}}(\omega_1)\cdot E^{\mathrm{loc}}(\omega_n),\end{aligned} \tag{5.29}$$

where (3.47) connects the number of elementary constituents per unit volume with the number of electrons per unit volume. On the other hand, nonlinear polarization can be expressed as result of the nonlinear coupling of an external electric field with an effective nonlinear susceptibility:

$$\boldsymbol{P}^{(n)}\left(\sum_{j=1}^{n}\omega_j\right) = \chi_{\text{eff}}^{(n)}\left(\sum_{j=1}^{n}\omega_j;\omega_1,\ldots,\omega_n\right) E(\omega_1)\cdot E(\omega_n). \tag{5.30}$$

In order to obtain an expression of the effective susceptibility, we have to equate expressions (5.29) and (5.30). Considering that the local field in a nonlinear medium can be expressed as

$$\boldsymbol{E}_{\text{loc}}(\omega) = \boldsymbol{E}(\omega) + \frac{4\pi}{3}\boldsymbol{P}^{(1)}(\omega) + \frac{4\pi}{3}\boldsymbol{P}^{\text{NL}}(\omega), \tag{5.31}$$

we find that the effective macroscopic nonlinear susceptibility, which takes into account local field effects, is related to the nth-order microscopic hyperpolarizability $\beta^{(n)}$ in the following way:

$$\begin{aligned}\chi_{\text{eff}}^{(n)}\left(\sum_{j=1}^{n}\omega_j;\omega_1,\ldots,\omega_n\right) &= \aleph L\left(\sum_{j=1}^{n}\omega_j\right) L(\omega_1)\ldots L(\omega_n)\\ &\quad\times\beta^{(n)}\left(\sum_{j=1}^{n}\omega_j;\omega_1,\ldots,\omega_n\right),\end{aligned} \tag{5.32}$$

where $L(\omega)$ is defined as:

$$\begin{aligned}L(\omega) &= \frac{\varepsilon_{\text{eff}}(\omega)+2}{3} = 1 + \frac{4\pi}{3}\chi_{\text{eff}}^{(1)}(\omega) = 1 + \frac{4\pi}{3}\frac{\aleph\alpha(\omega)}{1-\frac{4\pi}{3}\aleph\alpha(\omega)}\\ &= 1 + \frac{4\pi}{3}\frac{\chi^{(1)}(\omega)}{1-\frac{4\pi}{3}\chi^{(1)}(\omega)} = \frac{1}{1-\frac{4\pi}{3}\chi^{(1)}(\omega)},\end{aligned} \tag{5.33}$$

where we have used (3.55)–(3.56). Similarly to the linear case, if $\aleph$ is small,

$$\chi_{\text{eff}}^{(n)}\left(\sum_{j=1}^{n}\omega_j;\omega_1,\ldots,\omega_n\right) \sim \aleph\beta^{(n)}\left(\sum_{j=1}^{n}\omega_j;\omega_1,\ldots,\omega_n\right). \tag{5.34}$$

The expression of nonlinear susceptibility presented in this work has been derived from a strictly microscopic treatment derived from the nonlinear Green function presented in (5.26), so that

$$\chi^{(n)}\left(\sum_{j=1}^{n}\omega_j;\omega_1,\ldots,\omega_n\right) = \aleph\beta^{(n)}\left(\sum_{j=1}^{n}\omega_j;\omega_1,\ldots,\omega_n\right). \tag{5.35}$$

Hence, we find that the following relation holds between the effective macroscopic nonlinear susceptibility and the nonlinear susceptibility proposed in expression (5.27):

$$\chi_{\mathrm{eff}}^{(n)}\left(\sum_{j=1}^{n}\omega_j;\omega_1,\ldots,\omega_n\right) = L\left(\sum_{j=1}^{n}\omega_j\right) L(\omega_1)\ldots L(\omega_n) \times \chi^{(n)}\left(\sum_{j=1}^{n}\omega_j;\omega_1,\ldots,\omega_n\right). \tag{5.36}$$

The experimental setup for the measurement of linear and nonlinear local field factors was established by Maki et al. [45] who measured the reflectivity and surface phase conjugation of a dense atomic vapor. The former quantity was used to determine the linear microscopic polarizability and the latter for determination of the second hyperpolarizability. They found good agreement between the measured data and theoretical values calculated from the Clausius-Mossotti equation.

5.4.2 Two-Phase Media

In the field of optical physics, nanostructures have attracted much interest during the past two decades as a new class of nonlinear optical materials. There is a great demand for materials with large nonlinear optical responses and suitable spectral properties because it is assumed that such materials will play an important role in future photonic devices and all-optical signal processing [142, 143]. Conventionally, the development of a new optically nonlinear material has included a search for materials with suitable nonlinear optical properties. However, a more advanced way to enhance nonlinear properties would be to find suitable constituent materials and combine them into a nanostructure.

The nonlinear optical properties of nanostructures, including Maxwell Garnett [62, 144], Bruggeman [47, 71], metal-coated [145], and layered [48] nanostructures, have been studied widely both theoretically and experimentally [146, 147]. The interest in such nanocomposites arises from the possibility of enhancing their optical nonlinearity: the effective nonlinear optical susceptibility of the nanostructure material can exceed those of its constituent components in a wide range of nonlinear processes such as four-wave mixing, third-harmonic-generation, and nonlinear refractive index [148]. This enhancement is due to the collective enhancement of local electric fields in the medium. Recently, the enhancement of nonlinearity was verified experimentally [146]. Moreover, optical bistability, which has great potential in optical switching, has been demonstrated in weak nonlinear nanostructures [149] and even in a thin layer of optically nonlinear material [150], and has been explained in terms of local field effects.

The Maxwell Garnett Nanosphere System

The nonlinear optical properties of nanospheres [60,144,151–153] and coated nanospheres [145,154] have been intensively studied. The theory and mathematical formulas for effective third-order degenerate susceptibility were given by Sipe and Boyd [144] for cases where the nanospheres are optically nonlinear while the host material acts in an optically linear manner and vice versa. They also derived an expression for the general case where both the nanospheres and the host material are optically nonlinear. The effective degenerate third-order susceptibility for linearly polarized light can be expressed as follows (for a more detailed derivation see [144]):

$$\begin{aligned} \chi_{\mathrm{eff}}^{(3)}(\omega;\omega,\omega,-\omega) &= f_{\mathrm{i}}[L_{\mathrm{i}}(\omega)]^2|L_{\mathrm{i}}(\omega)|^2\chi_{\mathrm{i}}^{(3)}(-\omega;\omega,\omega,-\omega) \\ &+ [L_{\mathrm{h}}(\omega)]^2|L_{\mathrm{h}}(\omega)|^2[(1-f_{\mathrm{i}})+f_{\mathrm{i}}x(\omega)]\chi_{\mathrm{h}}^{(3)}(-\omega;\omega,\omega,-\omega), \end{aligned} \tag{5.37}$$

where the factor $x(\omega)$ is given by

$$x(\omega) = \frac{8}{5}[v(\omega)]^2|v(\omega)|^2 + \frac{6}{5}v(\omega)|v(\omega)|^2 + \frac{2}{5}[v(\omega)]^3 + \frac{18}{5}\left\{|v(\omega)|^2 + [v(\omega)]^2\right\}. \tag{5.38}$$

The local field factors of nanospheres (=inclusions) and the host material are, respectively,

$$L_{\mathrm{i}}(\omega) = \frac{\varepsilon_{\mathrm{eff}}(\omega) + 2\varepsilon_{\mathrm{h}}(\omega)}{\varepsilon_{\mathrm{i}}(\omega) + 2\varepsilon_{\mathrm{h}}(\omega)}, \tag{5.39}$$

$$L_{\mathrm{h}}(\omega) = \frac{\varepsilon_{\mathrm{eff}}(\omega) + 2\varepsilon_{\mathrm{h}}(\omega)}{3\varepsilon_{\mathrm{h}}(\omega)}. \tag{5.40}$$

The general expression (5.37) can be simplified if the volume fraction of the inclusions is small, as assumed in the MG model. The effective dielectric function of such a composite approaches the value of the host material and the local field factor of the host approaches unity. Now (5.37) can be written in the form

$$\begin{aligned} \chi_{\mathrm{eff}}^{(3)}(\omega;\omega,\omega,-\omega) &= f_{\mathrm{i}}[L_{\mathrm{i}}(\omega)]^2|L_{\mathrm{i}}(\omega)|^2\chi_{\mathrm{i}}^{(3)}(-\omega;\omega,\omega,-\omega) \\ &+ \chi_{\mathrm{h}}^{(3)}(-\omega;\omega,\omega,-\omega). \end{aligned} \tag{5.41}$$

The approximation of the effective third-order susceptibility is linearly dependent on the volume fraction f_{i} of the inclusions at small values of f_{i}. The recent study by Prot et al. [60] dealt with gold nanospheres embedded in a silica host. A large enhancement was observed in the imaginary part of the degenerate third-order nonlinear susceptibility, which was plotted as a function of the volume fraction of nanospheres. They observed that the linear dependence on f_{i} is lost when the volume fraction exceeds 10%, which is in agreement with the theory.

Equation (5.41) predicts the possibility of the cancellation of nonlinear absorption in a Maxwell Garnett nanosphere system. Smith et al. [153] measured the nonlinear absorption from a metal colloid of gold nanospheres that were embedded in the water, using the Z-scan technique. The cancellation is due mainly to the imaginary part of the local field factor of the inclusions. Materials with a large nonlinear refractive index and negligible nonlinear absorption are ideal, for instance, in optical switching. Unfortunately, in bulk materials, a high nonlinear refractive index is usually related to high nonlinear absorption. Therefore, the aim in nonlinear optical nanostructure engineering is usually to find optimum constituent materials, so that the nonlinear absorption of incident light is negligible. Conversely, in some cases, it may even be desirable to have large nonlinear absorption such as two-photon absorption-induced two-photon fluorescence from a microvolume, which has great potential in drug discovery and in novel bioaffinity assays [155]. In conclusion, the nonlinear optical properties of an MG nanosphere system can be optimized according to the applications by appropriate choice of the constituent materials and volume fraction.

The Bruggeman Effective Medium Theory

For randomly intermixed constituents, effective nonlinear susceptibilities should be considered with a statistical theory [59]. Fortunately, with the approximation of the uniformity of the electric field inside each constituent, the effective degenerate third-order susceptibility is given by [47]

$$\chi_{\mathrm{eff}}^{(3)}(\omega;\omega,\omega,-\omega)=\sum_i \frac{1}{f_i}\left|\frac{\partial\varepsilon_{\mathrm{eff}}(\omega)}{\partial\varepsilon_i(\omega)}\right|\left(\frac{\partial\varepsilon_{\mathrm{eff}}(\omega)}{\partial\varepsilon_i(\omega)}\right)\chi_i^{(3)}(-\omega;\omega,\omega,-\omega), \quad (5.42)$$

where i denotes the summation of the constituents. For a two-phase Bruggeman model, the explicit expressions for the derivatives have been presented in [47]. Equation (5.42) predicts that there can be a small enhancement in the third-order nonlinear susceptibility, when the linear refractive index of the linear constituent is larger than that of the nonlinear constituent [59]. In addition, Lakhtakia [156] investigated the complex linear dielectric function and nonlinear susceptibility of Bruggeman effective media using the strong permittivity fluctuation theory. One of the few experimental results was presented by Gehr et al. [64] who investigated the nonlinear properties of a porous glass saturated with optically nonlinear liquids by the Z-scan method. The enhancement of nonlinear susceptibility was verified and observed to be approximately 50%.

The Bruggeman formalism is applicable to nanosphere systems with shape distributions. Recently, it has been shown [157, 158] that shape variation can cause the separation of the nonlinear absorption peak and the enhanced nonlinear susceptibility peak. This means that it is possible to optimize the nonlinear enhancement of the nanostructure by finding an optimum shape distribution. However, the spectral properties of the constituents must be known.

Layered Nanostructures

A mathematical expression for the effective nonlinear susceptibility of layered nanostructures has been given by Boyd and Sipe [48] starting from the mesoscopic field inside each layer. The macroscopic field was obtained after an averaging process of the mesoscopic fields. For a two-phase structure with TE-polarized light, the effective degenerate third-order susceptibility becomes a summation of the constituent susceptibilities multiplied by their volume fractions, as follows:

$$\chi_{\mathrm{eff}}^{(3)}(\omega;\omega,\omega,-\omega) = f_{\mathrm{a}}\chi_{\mathrm{a}}^{(3)}(\omega;\omega,\omega,-\omega) + f_{\mathrm{b}}\chi_{\mathrm{b}}^{(3)}(\omega;\omega,\omega,-\omega). \tag{5.43}$$

Hence, their enhancement in a nonlinear optical process cannot exist. However, for TM-polarized light, the situation changes drastically. Boyd and Sipe [48] formulated the mathematical expression for an effective degenerate third-order susceptibility with TM-polarized light:

$$\begin{aligned}\chi_{\mathrm{eff}}^{(3)}(\omega;\omega,\omega,-\omega) = {} & f_{\mathrm{a}}\left|\frac{\varepsilon_{\mathrm{eff}}(\omega)}{\varepsilon_{\mathrm{a}}(\omega)}\right|^2\left[\frac{\varepsilon_{\mathrm{eff}}(\omega)}{\varepsilon_{\mathrm{a}}(\omega)}\right]^2\chi_{\mathrm{a}}^{(3)}(\omega;\omega,\omega,-\omega)\\ & + f_{\mathrm{b}}\left|\frac{\varepsilon_{\mathrm{eff}}(\omega)}{\varepsilon_{\mathrm{b}}(\omega)}\right|^2\left[\frac{\varepsilon_{\mathrm{eff}}(\omega)}{\varepsilon_{\mathrm{b}}(\omega)}\right]^2\chi_{\mathrm{b}}^{(3)}(\omega;\omega,\omega,-\omega).\end{aligned} \tag{5.44}$$

Therefore, if $\varepsilon_{\mathrm{a}} > \varepsilon_{\mathrm{b}}$ and $\chi_{\mathrm{a}}^{(3)} < \chi_{\mathrm{b}}^{(3)}$, the effective third-order susceptibility exceeds those of its constituent components. The terms in brackets in (5.44) represent the local field factors for the constituents. The analysis of the nonlinear properties of layered structures presented above is based on the effective medium approximation. Furthermore, an analysis based on a nonlinear wave equation has been examined [159], yielding analogous results. Theoretically, there can be enhancement up to a factor of 10 when the ratio of the linear refractive indices of the constituents is equal to 2. In their experimental measurement, Fischer et al. [146] obtained enhancement only by a factor of 1.35. This was caused by the small difference between the linear refractive indices of the organic polymer investigated and of titanium dioxide, in agreement with the theory.

Layered nanostructures are assumed to play an important role in the frame of nonlinear optics. It has been demonstrated that the electro-optic effect can be enhanced with layered nanostructures [160]. The enhancement of nonlinearity in layered structures makes them very attractive in all-optical signal processing. Kishida et al. [142, 143] observed large nonlinearity in transition-metal oxides and halides. Such materials combined into a layered nanostructure may possess huge third-order nonlinearity and may find applications in novel optoelectronic device development.

5.4.3 Tailoring of the Optical Properties of Nanostructures

In this section, the linear and nonlinear optical properties of Maxwell Garnett, Bruggeman, and layered nanostructures are simulated using two conjugated polymers (polythiophene PT10 and polysilane PDHS) with nanoscale TiO_2 particles. Polythiophenes as well as polydiacetylenes are π-conjugated organic polymers, whereas polysilanes are σ-conjugated polymers of Si atoms. Most of the conjugated polymers have a center of symmetry, and their optical nonlinearities are dominated by third-order susceptibilities. These (as well as linear susceptibilities) can be described by the properties of excitons (electron–hole pairs) confined in one-dimensional (1-D) geometry [161, 162]. Although conjugated polymers have relatively large nonlinear optical susceptibilities, their magnitudes may still fall short of that needed for many proposed applications in photonics.

The present simulations are performed for an effective medium using the equations presented in the previous sections. Since the effective nonlinear susceptibility of a nanostructure is assumed to possess simultaneously real and imaginary parts, the simulations are based on the use of the modulus of the effective nonlinear susceptibility. Actually, in nonlinear optics, many practical measurements yield data on the modulus of nonlinear susceptibility. For the three effective medium models, the nonlinear material is a conjugated polymer. It is either a polythiophene, poly(3-decylthiophene) (PT10), or a polysilane, poly(dihexylsilane) (PDHS), both exhibit a third-order nonlinear optical response. The linear material is amorphous TiO_2, which has a high linear refractive index and negligible optical nonlinearity. Both polymers have a linear refractive index lower than that of TiO_2, as presented in Fig. 5.3a, predicting the enhancement in nonlinearity.

The dielectric function for amorphous TiO_2 was obtained from the ellipsometric study of Joseph and Gagnaire [164]. For PT10 and PDHS, the dielectric functions were computed by the Kramers-Kronig analysis with the aid of published absorption spectra [165, 166] and refractive index data [167, 168]. The degenerate third-order susceptibilities of PT10 and PDHS were calculated by using the standard formulas given in [169]. The dipole matrix elements needed in the calculation were obtained from the experimental analysis by Torruellas et al. [166] (in the case of PT10) and Hasegawa et al. [165] (PDHS). Figure 5.3b shows the amplitude spectrum of $\chi^{(3)}(-\omega;\omega,\omega,-\omega)$ of polymers PT10 and PDHS. The resonance peak of PT10 at 2.4 eV is almost twice as large as that of PDHS at 3.3 eV. The effective dielectric function and the effective third-order nonlinear susceptibility were first calculated for nanostructures. Figure 5.4a,b shows the modulus of the corresponding effective nonlinear susceptibilities. The modulus of the effective third-order nonlinear susceptibility was used to calculate the enhancement of the composite as a function of the volume fraction of the inclusions, as follows:

$$\text{Enhancement} = \frac{|\chi^{(3)}_{\text{eff}}(\omega;\omega,\omega,-\omega)|_{\max}}{|\chi^{(3)}_{\text{a}}(\omega;\omega,\omega,-\omega)|_{\max}}, \tag{5.45}$$

where the maximum of the modulus is obtained from spectral data shown in Fig. 5.4. In (5.45) a denotes the constituent nonlinear polymer PT10 or

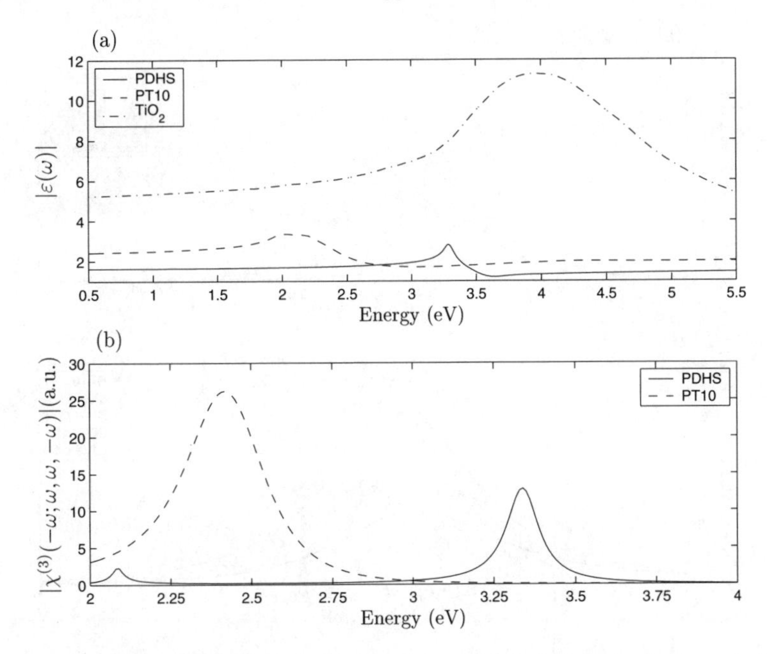

Fig. 5.3. Amplitude spectra of the modulus of (**a**) dielectric function and (**b**) nonlinear degenerate third-order susceptibility of the constituent materials. Reproduced from [163]

PDHS. The largest enhancement is obtained for layered nanostructures, as presented in Figs. 5.5a–5.8a. The upper limit of volume fraction was restricted to a value of 0.25 in the MG nanosphere system. From Fig. 5.5a, it can be observed that the enhancement in the MG model is a monotonically growing function of the volume fraction of the present constituents. However, in Bruggeman and layered structures there will be a maximum value for the enhancement, as presented in Figs. 5.6a–5.8a. Moreover, the enhancement in such structures is an unambiguous function of the volume fraction. In other words, for instance, in the case of Fig. 5.7a, the same enhancement (= 1.75) can be obtained when the volume of the inclusions is either approximately 0.35 or 0.62. This indicates that in Bruggeman and layered nanostructures, and in the frame of the present constituents, it is possible to optimize the enhancement by choosing the appropriate volume fractions of constituents.

As shown in Sect. 3.4.2, in the MG nanostructures, the maximum of the linear absorption spectrum shifts as a function of the volume fraction of inclusions. This shift occurs usually toward the lower energies, as the volume fraction of inclusions increases [62]. In the nonlinear case, it is possible to evaluate the blueshift of the peak of the modulus of the degenerate nonlinear susceptibility of nanostructures as a function of the volume fraction of the TiO_2 constituent. The relative blueshift is defined as follows:

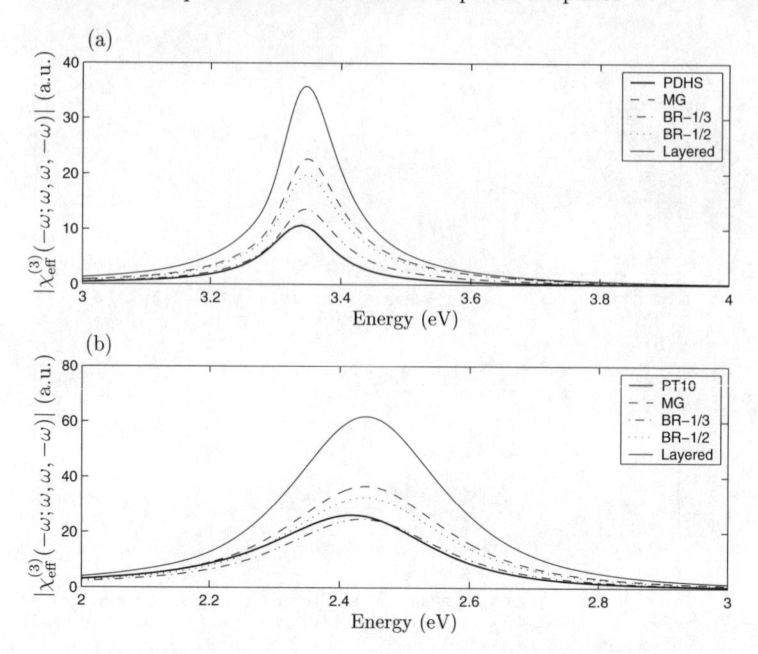

Fig. 5.4. Amplitude spectra of the modulus of the effective nonlinear susceptibility of (**a**) PDHS-TiO_2 and (**b**) PT10-TiO_2 nanostructures. *Bold lines*: pure polymers. *Dashed lines*: Maxwell Garnett nanosphere system with volume fraction $f_{TiO_2} = 0.25$. *Dash-dotted lines*: Bruggeman model with shape factor $g = 1/3$ and volume fraction $f_{TiO_2} = 0.5$. *Dotted lines*: Bruggeman model with shape factor $g = 1/2$ and volume fraction $f_{TiO_2} = 0.5$. *Solid lines*: Layered nanostructure with volume fraction $f_{TiO2} = 0.5$. Reproduced from [163]

$$\mathrm{BS} = \frac{E_{\mathrm{rp-eff}} - E_{\mathrm{rp-0}}}{E_{\mathrm{rp-0}}} \times 100\%, \tag{5.46}$$

where $E_{\mathrm{rp-eff}}$ denotes the energy of the resonance of the modulus of the effective nonlinear susceptibility of the nanostructure and $E_{\mathrm{rp-0}}$ is the corresponding energy of the resonance of the modulus of the nonlinear susceptibility of the pure polymer itself (for PT10 $\approx 2.4\,\mathrm{eV}$ and for PDHS $\approx 3.3\,\mathrm{eV}$). The results of the calculations are presented in Figs. 5.5b–5.8b. It can be observed that for all investigated topologies, a larger enhancement is obtained with PDHS-TiO_2 nanostructures than with PT10-TiO_2 constituents. On the contrary, the relative blueshift of a PDHS-TiO_2 nanostructure is always lower than that of PT10-TiO_2. In addition, in the MG model, the relative blueshift is practically a linear function of the volume fraction. However, the relative blueshift is a nonlinear function of the volume fraction in Bruggeman and layered nanostructures. The volume fraction that gives the maximum relative blueshift in such topology models can be observed from Figs. 5.6b–5.8b.

In conclusion, the simulations show that it is possible to tailor the nonlinear optical properties of nanostructures by employing the frequency de-

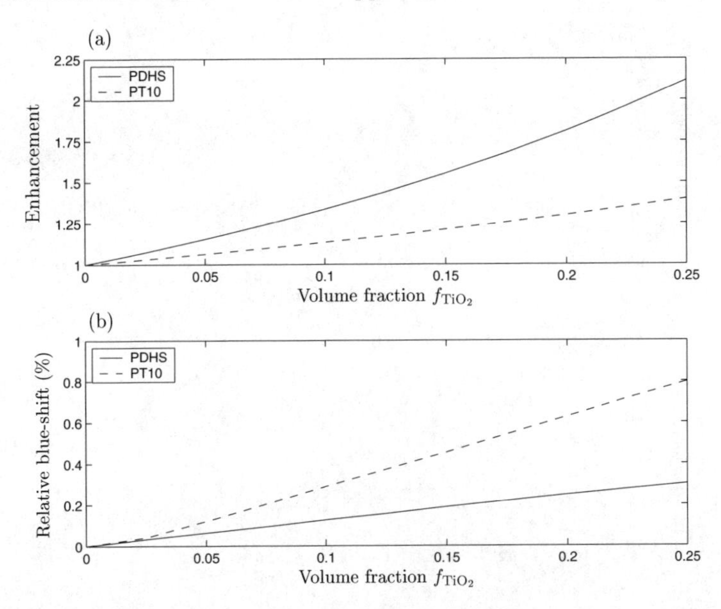

Fig. 5.5. (**a**) The enhancement and (**b**) relative blueshift of the effective degenerate susceptibility of a Maxwell Garnett nanosphere system. Reproduced from [163]

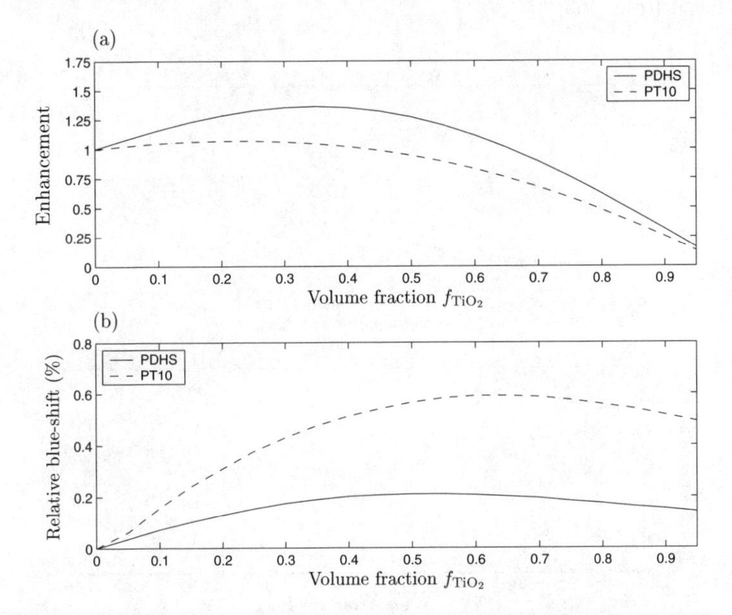

Fig. 5.6. (**a**) The enhancement and (**b**) relative blueshift of the effective degenerate susceptibility of the Bruggeman model with a geometric shape factor $g = 1/3$. Reproduced from [163]

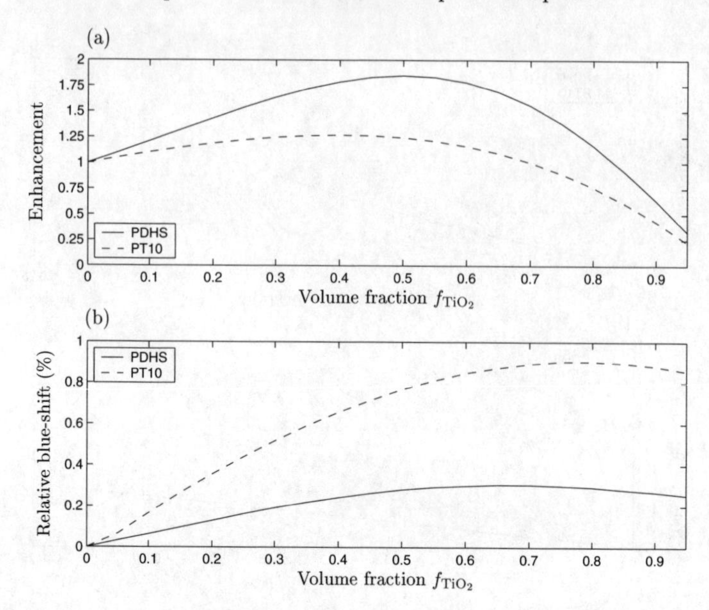

Fig. 5.7. (**a**) The enhancement and (**b**) relative blueshift of the effective degenerate susceptibility of the Bruggeman model with a geometric shape factor $g = 1/2$. Reproduced from [163]

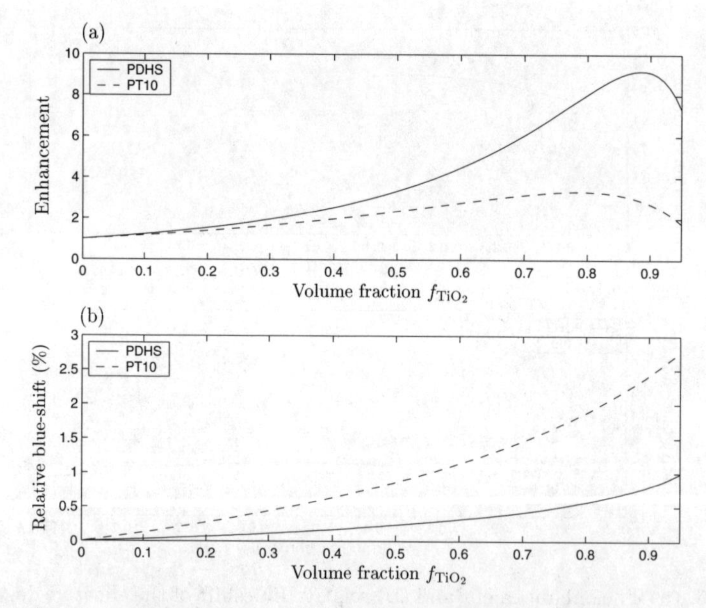

Fig. 5.8. (**a**) The enhancement and (**b**) relative blueshift of the effective degenerate susceptibility of a layered nanostructure. Reproduced from [163]

pendency of the linear and nonlinear susceptibilities of the constituents. In particular, it is possible to tune both the enhancement and relative blueshift simply by tuning the volume fraction of the constituents of the nanostructure or by choosing the appropriate topology of the nanostructure. In principle, the enhancement of nonlinear susceptibility due to the self-action process provides a method for changing the intensity-dependent refractive index of the nanostructure. This enhancement can be used, for instance, in optical switching. In turn, the relative blueshift provides a method for wavelength multiplexing.

6 Kramers-Kronig Relations and Sum Rules in Nonlinear Optics

6.1 Introductory Remarks

Research in nonlinear optics has usually focused on achieving high resolution in both experimental data and theoretical calculations. On the contrary, in spite of the ever increasing scientific and technological relevance of nonlinear optical phenomena, relatively little attention has been paid to the experimental investigation of K-K dispersion relations and the sum rules of the corresponding nonlinear susceptibilities. These properties are especially relevant for experimental investigations of frequency-dependent nonlinear optical properties. In the context of this kind of analysis, K-K relations and sum rules could provide information whether or not a coherent, common picture of the nonlinear properties of the material under investigation is available. We remark that only few data sets referring to nonlinear optical phenomena span a spectral range wide enough to permit the usage of dispersion relations.

The first heuristic applications of K-K dispersion relations theory to nonlinear susceptibilities date back to the 1960s [170–172] and 1970s [173, 174], while a more systematic study began within the last decade. Some authors have preferentially introduced the K-K relations in the context of ab initio or model calculations of materials properties [175–179]; a complete review of this approach can be found in [26]. Other authors have used a more general approach capable of providing the theoretical foundations of dispersion theory for nonlinear optics [20–23, 25, 180]. These permit a connection between K-K relations and the establishment of sum rules for the real and imaginary parts of susceptibility. The instruments of complex analysis permit the definition of the necessary and sufficient conditions for the applicability of K-K relations, which require that the nonlinear susceptibility function descriptive of the nonlinear phenomena under examination is holomorphic in the upper complex plane of the relevant frequency variable. The asymptotic behavior of nonlinear susceptibility determines the number of independent pairs of K-K relations that hold simultaneously. Combining K-K relations and the knowledge of the asymptotic behavior of nonlinear susceptibility, it is possible to derive, along the same lines as in the linear case, sum rules for nonlinear optics. Sum rules for nonlinear optics can also be obtained using approaches relying heavily on statistical physics methods [181]. A comprehensive analysis of Kramers-Kronig relations and sum rules in nonlinear optics can be found in [34].

We emphasize that in nonlinear optics, similarly to the linear case, when both the real and imaginary parts of holomorphic susceptibilities can be experimentally measured as independent quantities, K-K relations provide the means to estimate the self-consistency of measured data. Furthermore, generalized K-K relations as well as multiply subtractive K-K relations provide also in such a case the means to judge the success of measured data.

In this chapter, we study the holomorphic properties of general nonlinear susceptibility functions, and we deduce general criteria for establishing the cases in which K-K relations connect the real and imaginary parts of nonlinear susceptibility under consideration. We will also consider the analysis of the nonlinear integral properties of pump-and-probe susceptibility as an experimentally and theoretically relevant example.

6.2 Kramers-Kronig Relations in Nonlinear Optics: Independent Variables

The nonlinear Green function obeys causality for each of its time variables, as shown in (5.4). Therefore, assuming as usual that suitable integrability conditions are obeyed, we can apply the Titchmarsch theorem [2] separately to each variable and deduce that the nonlinear susceptibility function (5.7) is holomorphic in the upper complex plane of each variable ω_i, $1 \leq i \leq n$. If we consider the first argument ω_1 of nonlinear susceptibility function (5.7), the following relation holds

$$\mathrm{P}\int_{-\infty}^{\infty} \frac{\chi^{(n)}_{ij_1\ldots j_n}\left(\sum_{l=1}^{n}\omega'_l;\omega'_1,\ldots,\omega'_n\right)}{\omega'_1-\omega_1}\mathrm{d}\omega'_1 = \mathrm{i}\pi\chi^{(n)}_{ij_1\ldots j_n}\left(\sum_{l=1}^{n}\omega_l;\omega_1,\ldots,\omega'_n\right). \tag{6.1}$$

Repeating the same procedure for all remaining $(n-1)$ frequency variables and applying the symmetry relation (5.10), we obtain

$$\begin{aligned}&\left(-\frac{\pi}{2}\right)^n \mathrm{Im}\left\{\chi^{(n)}_{ij_1\ldots j_n}\left(\sum_{l=1}^{n}\omega_l;\omega_1,\ldots,\omega_n\right)\right\}\\ &= \omega_1\ldots\omega_n \mathrm{P}\int_0^{\infty}\ldots\mathrm{P}\int_0^{\infty}\frac{\mathrm{Re}\left\{\chi^{(n)}_{ij_1\ldots j_n}\left(\sum_{l=1}^{n}\omega'_l;\omega'_1,\ldots,\omega'_n\right)\right\}}{({\omega'_1}^2-\omega_1^2)\ldots({\omega'_n}^2-\omega_n^2)}\mathrm{d}\omega'_1\ldots\mathrm{d}\omega'_n\end{aligned} \tag{6.2}$$

and

$$\left(\frac{\pi}{2}\right)^n \mathrm{Re}\left\{\chi^{(n)}_{ij_1\ldots j_n}\left(\sum_{l=1}^{n}\omega_l;\omega_1,\ldots,\omega_n\right)\right\}$$
$$= \mathrm{P}\int_0^\infty \ldots \mathrm{P}\int_0^\infty \frac{\omega_1'\ldots\omega_n'\mathrm{Im}\left\{\chi^{(n)}_{ij_1\ldots j_n}\left(\sum_{l=1}^{n}\omega_l';\omega_1',\ldots,\omega_n'\right)\right\}}{({\omega_1'}^2-\omega_1^2)\ldots({\omega_n'}^2-\omega_n^2)}\mathrm{d}\omega_1'\ldots\mathrm{d}\omega_n' \tag{6.3}$$

which for nonlinear optics are the equivalent of the conventional K-K relations described for the linear case in Sect. 4.3. It is clear that these relations, in spite of their theoretical significance, are not interesting from an experimental point of view, since their verification as well as their utilization for optical data retrieving requires the possibility of independently changing n laser beams. Moreover, most of the physically relevant nonlinear phenomena are described by nonlinear susceptibilities where all or part of the frequency variables are mutually dependent. This occurs in any case when the number of the frequency components of the incoming radiation is less than the order of the nonlinear process under examination, as occurs in the pump-and-probe case (5.19). Dispersion relations (6.2) and (6.3) are then a mere mathematical extension of linear K-K relations. We may, therefore, understand that a more flexible theory, where the nonlinearity of the interaction is fully acknowledged, is needed in order to provide effectively relevant dispersion relations for nonlinear optical phenomena.

6.3 Scandolo's Theorem and Kramers-Kronig Relations in Nonlinear Optics

In nonlinear optics, a relevant dispersion relation is a line integral in the space of frequency variables, which entails the choice of a one-dimensional space embedded in an n-dimensional space, if we consider an nth-order nonlinear process. This corresponds to realistic experimental setting where only the frequency of one of the monochromatic beams described in (5.12) is changed. Since we have frequency mixing in nonlinear optics, changing one frequency of the incoming radiation will change none, one, or more than one arguments of the nonlinear susceptibility function considered, depending on how many photons of the tunable incoming radiation it takes into account. The choice of the parametrization then selects different susceptibilities and so refers to different nonlinear optical processes. Each component j of the straight line in $\mathbb{R}^n$ can be parametrized as follows:

$$\omega_j(s) = v_j s + w_j \quad , \quad 1 \le j \le n, \tag{6.4}$$

where the parameter $s \in (-\infty,\infty)$, the vector $\boldsymbol{v}$ of its coefficients describes the direction of the straight line, and the vector $\boldsymbol{w}$ determines $\boldsymbol{\omega}(0)$.

We refer to the previously presented pump-and-probe susceptibility (5.19) to provide well-suited examples of the parametrizations (6.4). In this case, while the pump frequency ω_2 is fixed, the probe frequency ω_1 can be tuned. The correct parametrization of the straight line for the first term in expression (5.19) is then $\boldsymbol{v} = (1, 0, 0)$ and $\boldsymbol{w} = (0, \omega_2, -\omega_2)$. The correct parametrization of the second term in (5.19) is $\boldsymbol{v} = (1, 1, -1)$ and $\boldsymbol{w} = (0, 0, 0)$, because all arguments change simultaneously but one has a sign opposite to the other two.

Scandolo's theorem [23] permits the determination in very general terms of the holomorphic properties of the nonlinear susceptibility tensor with respect to the varying parameter s. Substituting expression (6.4) in definition (5.7), we obtain

$$\begin{aligned} &\chi^{(n)}_{ij_1 \dots j_n}\left[\sum_{j=1}^{n}(v_j s + w_j); v_1 s + w_1, \dots, v_n s + w_n\right] \\ &= \int_{-\infty}^{\infty} G^{(n)}_{ij_1 \dots j_n}(t_1, \dots, t_n) \exp\left(\mathrm{i}s\sum_{j=1}^{n} v_j t_j + \mathrm{i}\sum_{j=1}^{n} w_j t_j\right) \mathrm{d}t_1 \dots \mathrm{d}t_n . \end{aligned} \tag{6.5}$$

When dealing with the s-parametrized form of nonlinear susceptibility, from now on we will use the simpler notation

$$\chi^{(n)}_{ij_1 \dots j_n}(s) = \chi^{(n)}_{ij_1 \dots j_n}\left[\sum_{j=1}^{n}(v_j s + w_j); v_1 s + w_1, \dots, v_n s + w_n\right] . \tag{6.6}$$

We compute the Fourier inverse transform with respect to s of expression (6.5) and obtain the following parametrized one-variable nonlinear Green function:

$$G^{(n)}_{ij_1 \dots j_n}(\tau) = \frac{1}{2\pi}\int_{-\infty}^{\infty} \chi^{(n)}_{ij_1 \dots j_n}(s)\, e^{-\mathrm{i}s\tau} \mathrm{d}s. \tag{6.7}$$

Under the condition

$$v_j \geq 0, \quad 1 \leq j \leq n, \tag{6.8}$$

we deduce that, given that the nonlinear Green function obeys causality for each time variable, as remarked in condition (5.4), expression (6.7) obeys the following condition with respect to the variable τ:

$$G^{(n)}_{ij_1 \dots j_n}(\tau) = 0, \quad \tau \leq 0. \tag{6.9}$$

With (6.9) and making the usual assumption that the function belongs to the space L^2, we can take advantage of Titchmarsch theorem [2] and deduce that the nonlinear susceptibility (6.6) is holomorphic in the upper complex plane of the complex variable s. Hence, the Hilbert transforms connect the real and imaginary parts of the nonlinear susceptibility (6.6):

$$\mathrm{Re}\left\{\chi^{(n)}_{ij_1\ldots j_n}(s)\right\} = \frac{1}{\pi}\mathrm{P}\int_{-\infty}^{\infty}\frac{\mathrm{Im}\left\{\chi^{(n)}_{ij_1\ldots j_n}(s')\right\}}{s'-s}\mathrm{d}s' \tag{6.10}$$

$$\mathrm{Im}\left\{\chi^{(n)}_{ij_1\ldots j_n}(s)\right\} = -\frac{1}{\pi}\mathrm{P}\int_{-\infty}^{\infty}\frac{\mathrm{Re}\left\{\chi^{(n)}_{ij_1\ldots j_n}(s')\right\}}{s'-s}\mathrm{d}s'. \tag{6.11}$$

These general relations extend the results obtained in the 1960s [170–172] and 1970s [173,174] for specific nonlinear phenomena, and the general conclusions drawn in the 1980s for specific models [20–22]. The condition (6.8) on the sign of the directional vectors of the straight line in $\mathbb{R}^n$ implies that only one particular class of nonlinear susceptibilities possesses the holomorphic properties required to obey dispersion relations (6.10) and (6.11). Hence, causality is not a sufficient condition for the existence of K-K relations between the real and imaginary parts of a general nonlinear susceptibility function. The principle of causality [3] of the response function is reflected mathematically in the validity of the K-K relations for the nonlinear susceptibility in the form presented in (6.2) and (6.3).

We observe that the Scandolo theorem implies that the two terms of the pump-and-probe susceptibility (5.19) have fundamentally different properties.

The first term is holomorphic in the upper complex ω_1-plane, so that by considering the parametrization $\boldsymbol{v} = (1,0,0)$ and $\boldsymbol{w} = (0,\omega_2,-\omega_2)$ for the line integral, K-K relations can be established between the real and imaginary parts:

$$\mathrm{Re}\left\{\chi^{(3)}_{ij_1j_2j_3}(\omega_1;\omega_1,\omega_2,-\omega_2)\right\} = \frac{2}{\pi}\mathrm{P}\int_0^{\infty}\frac{\omega_1'\mathrm{Im}\left\{\chi^{(3)}_{ij_1j_2j_3}(\omega_1';\omega_1',\omega_2,-\omega_2)\right\}}{\omega_1'^2-\omega_1^2}\mathrm{d}\omega_1', \tag{6.12}$$

$$\mathrm{Im}\left\{\chi^{(3)}_{ij_1j_2j_3}(\omega_1;\omega_1,\omega_2,-\omega_2)\right\} = -\frac{2\omega_1}{\pi}\mathrm{P}\int_0^{\infty}\frac{\mathrm{Re}\left\{\chi^{(3)}_{ij_1j_2j_3}(\omega_1';\omega_1',\omega_2,-\omega_2)\right\}}{\omega_1'^2-\omega_1^2}\mathrm{d}\omega_1', \tag{6.13}$$

where we have considered the general tensorial component and have presented the usual semi-infinite integration by taking advantage of the relation (5.8). Similar dispersion relations can be established for the dominant term of the nonlinear change in the index of refraction presented in expression (5.20). Figure 6.1 shows the results of a study [27] where the K-K relations for the nonlinear change in the index of refraction were used heuristically before they had been rigorously derived.

The second term of (5.19) is not holomorphic in the upper complex ω_1-plane. The optical data related to the degenerate susceptibility cannot be inverted using K-K relations [71].

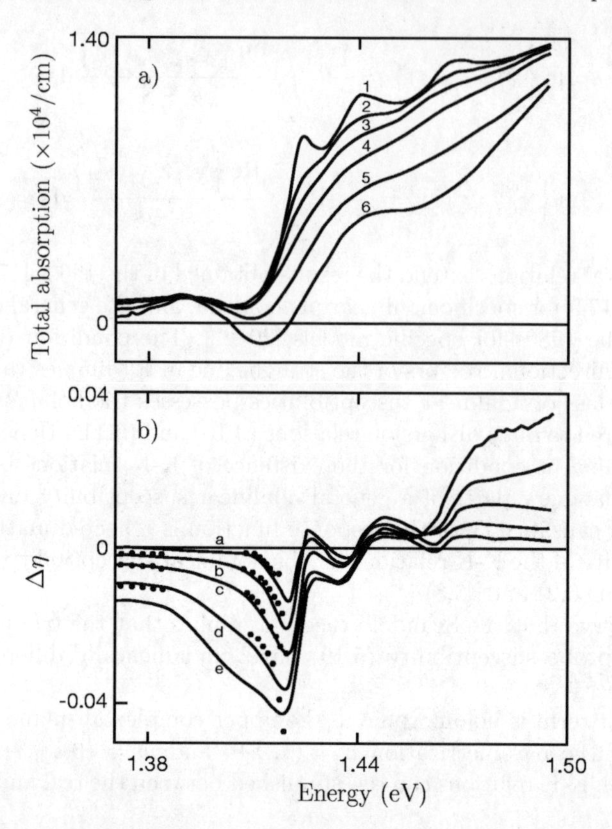

Fig. 6.1. Application of K-K relations for the inversion of nonlinear optical data in pump-and-probe experiments: (**a**) absorption at the probe frequency for different intensities of the pump laser; (**b**) *dots*: experimental data showing the real part of the nonlinear refractive index; *solid lines*: corresponding curves computed via K-K relations from the absorption data in (**a**). Reproduced from [27]

Since the first term in the conventional experimental setting is larger than the second term by orders of magnitude because of condition (5.18), it is reasonable to expect that the full pump-and-probe susceptibility (5.19) obeys K-K relations with good approximation.

For susceptibility functions that do not obey the conditions of the Scandolo theorem, i.e. if at least one component of the vector $\boldsymbol{v}$ in the parametrization (6.4) is negative, it is not possible to invert the optical data through K-K relations. Considering that these functions have poles in the upper complex s-plane, Cauchy's integral theorem [39] permits us to write the following relation:

$$\mathrm{P}\int_{-\infty}^{\infty} \frac{\chi^{(3)}_{ij_1\ldots j_n}(s)}{s'-s}\mathrm{d}s' = \mathrm{i}\pi\chi^{(n)}_{ij_1\ldots j_n}(s) + 2\pi\mathrm{i}\sum_{\mathrm{Im}(\Omega)>0}^{\mathrm{poles}} \mathrm{Res}\left[\frac{\chi^{(n)}_{ij_1\ldots j_n}(\Omega)}{\Omega - s}\right], \quad (6.14)$$

where the sum in the second term on the right hand-side member is over the residua (Res) of the susceptibility in the poles p_s in the upper complex s-plane. Property (5.9) implies that all the poles are symmetrical with respect to the imaginary axis of the variable s. Therefore, for these susceptibility functions, data inversion cannot be achieved unless, unrealistically, information on all the poles is provided. A recent study of a simplified oscillator model [129] shows that the knowledge of the poles in the upper complex ω_1-plane, resulting from the contribution to the total pump-and-probe susceptibility relative to the dynamical Kerr effect, permits us to write explicitly a set of dispersion relations between the real and the imaginary parts, which differ from the usual K-K relations by a term that is proportional to the intensity of the probe beam. Generally, the inversion of optical data for this class of susceptibility functions can be performed with a high degree of accuracy using algorithms based on the maximum entropy method [182], which in recent studies has shown its potentially great impact on applications [15,71,183–186]. We will deal with the maximum entropy method later in this book.

6.4 Kramers-Kronig Analysis of the Pump-and-Probe System

We consider the response of a physical system to a pump-and-probe experiment in the limit of very low intensity of the probe laser. We then consider the contribution to nonlinear polarization corresponding to the first term in expression (5.19). We can deduce the asymptotic behavior of this function, which is holomorphic in the upper complex ω_1-plane [23–25,34], so that the following relation holds for large values of ω_1:

$$\chi^{(3)}_{ij_1j_2j_3}(\omega_1;\omega_1,\omega_2,-\omega_2) \approx \frac{c_{ij_1j_2j_3}(\omega_2)}{\omega_1^4}. \tag{6.15}$$

As with the linear case, the leading asymptotic term of susceptibility for large values of ω_1 is real. The numerical evaluation of the ω_2-dependent tensor appearing as the coefficient of the leading asymptotic order has been performed in the case of the hydrogen atom [24]. From asymptotic equivalence (6.15), we deduce that the second moment of the nonlinear susceptibility considered here decreases asymptotically as ω_1^{-2} and that it is holomorphic in the upper complex ω_1-plane because it is a product of two holomorphic functions. We then deduce that the following pair of independent K-K relations holds [23]:

$$\mathrm{Re}\left\{\chi^{(3)}_{ij_1j_2j_3}(\omega_1;\omega_1,\omega_2,-\omega_2)\right\} = \frac{2}{\pi\omega_1^2}\mathrm{P}\int_0^\infty \frac{{\omega_1'}^3\mathrm{Im}\left\{\chi^{(3)}_{ij_1j_2j_3}(\omega_1';\omega_1',\omega_2,-\omega_2)\right\}}{{\omega_1'}^2-\omega_1^2}\mathrm{d}\omega_1', \tag{6.16}$$

$$\mathrm{Im}\left\{\chi^{(3)}_{ij_1j_2j_3}(\omega_1;\omega_1,\omega_2,-\omega_2)\right\} = -\frac{2}{\pi\omega_1}\mathrm{P}\int_0^\infty \frac{{\omega_1'}^2\mathrm{Re}\left\{\chi^{(3)}_{ij_1j_2j_3}(\omega_1';\omega_1',\omega_2,-\omega_2)\right\}}{{\omega_1'}^2-\omega_1^2}\mathrm{d}\omega_1'. \tag{6.17}$$

This pair of K-K relations is peculiar to the nonlinear response at the probe frequency because they are independent of the previously obtained dispersion relations (6.12) and (6.13). This in turn implies that the experimental or model-based data of the susceptibility considered have to obey *both* dispersion relations pairs in order to be self-consistent. We emphasize that similar dispersion relations also hold for the dominant part of the nonlinear contribution to the index of refraction presented in expression (5.20).

By applying the superconvergence theorem [9] presented in the previous chapter to K-K relations (6.12)–(6.17) and taking into account asymptotic behavior (6.15), it is possible to obtain sum rules for the moments of the real and imaginary parts of the susceptibility considered [23]:

$$\int_0^\infty \mathrm{Re}\left\{\chi^{(3)}_{ij_1j_2j_3}(\omega_1';\omega_1',\omega_2,-\omega_2)\right\}\mathrm{d}\omega_1' = 0, \tag{6.18}$$

$$\int_0^\infty \omega'^2\mathrm{Re}\left\{\chi^{(3)}_{ij_1j_2j_3}(\omega_1';\omega_1',\omega_2,-\omega_2)\right\}\mathrm{d}\omega_1' = 0, \tag{6.19}$$

$$\int_0^\infty \omega'\mathrm{Im}\left\{\chi^{(3)}_{ij_1j_2j_3}(\omega_1';\omega_1',\omega_2,-\omega_2)\right\}\mathrm{d}\omega_1' = 0, \tag{6.20}$$

$$\int_0^\infty \omega'^3\mathrm{Im}\left\{\chi^{(3)}_{ij_1j_2j_3}(\omega_1';\omega_1',\omega_2,-\omega_2)\right\}\mathrm{d}\omega_1' = -\frac{\pi}{2}c_{ij_1j_2j_3}(\omega_2). \tag{6.21}$$

These sum rules are a priori constraints that every set of experimental or model generated data has to obey simultaneously.

Sum rule (6.18) implies that the average of the real part of the nonlinear susceptibility is zero, as for the real part of linear susceptibility.

Sum rule (6.19), which applies only in the nonlinear case, given that the second moment of the real part of the linear susceptibility does not converge, gives an additional constraint on the negative and positive contributions in the real part of the nonlinear susceptibility.

Sum rule (6.20) implies that the imaginary part of the nonlinear susceptibility at the probe frequency has to be negative in some parts of the spectrum, so that nonlinear stimulated emissions compensate for nonlinear stimulated absorptions such as two-photon and stimulated Raman. In 1997, Cataliotti et al. [30] experimentally verified this sum rule on cold cesium atoms ($T < 10\,\mu\mathrm{K}$). When the pump laser is opportunely tuned, there is a

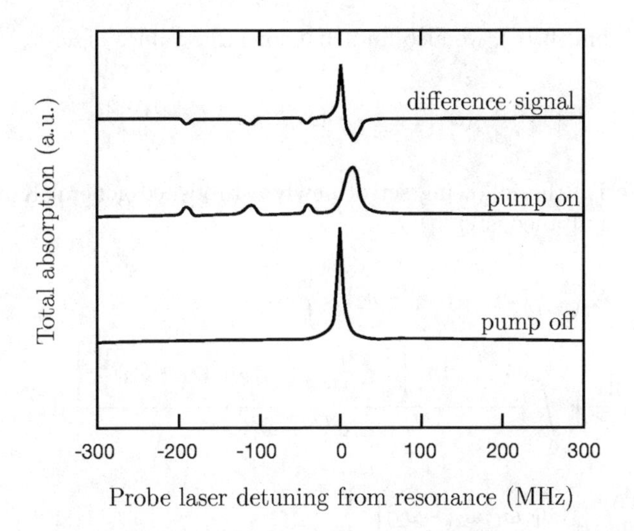

Fig. 6.2. *Bottom graph*: linear absorption; *Middle graph*: total absorption with pump laser switched on in EIT configuration; *Top graph*: nonlinear absorption. Reproduced from [30]

decrease in the main resonance absorption peak from quantum interference effects, as prescribed by electromagnetically induced transparency (EIT) theory [187], while new peaks of absorption outside the linear resonance are formed, as can be seen in Fig. 6.2. Those peaks compensate for the reduced absorption resulting from the EIT effect, so that the integral of the nonlinear absorption is zero with good experimental precision.

Sum rule (6.21) is peculiar to nonlinear susceptibility and links the absorption spectrum to a function only of the pump laser frequency and the specific material thermodynamic equilibrium density matrix.

Experimental analysis of the second and fourth sum rules would be of particular relevance because these constraints hold only in the nonlinear case. Nevertheless, we can expect that the actual verification of the sum rules may not be easily achievable because, in general, sum rules are integral relations whose validification requires data over very large spectral ranges [12, 75, 76].

6.4.1 Generalization of Kramers-Kronig Relations and Sum Rules

Along the lines of that presented in [78] for linear optics and in [188] for nonlinear phenomena for the specific case of harmonic-generation susceptibility, we propose an extension of the previously derived K-K relations (6.12)–(6.17) and sum rules (6.18)–(6.21) for the holomorphic contribution to pump-and-probe susceptibility by considering higher powers of the susceptibility. The kth ($k \geq 1$) power of the pump-and-probe susceptibility is holomorphic in the upper complex ω-plane, since it is the product of functions that obey this property. From (6.15), we also find that for large values of ω,

$$\left[\chi^{(3)}_{ij_1j_2j_3}(\omega_1;\omega_1,\omega_2,-\omega_2)\right]^k \approx \frac{[c_{ij_1j_2j_3}(\omega_2)]^k}{\omega_1^{4k}}. \tag{6.22}$$

We thus derive the following set of newly established general K-K relations that hold simultaneously:

$$\begin{aligned}&\mathrm{Re}\left\{\left[\chi^{(3)}_{ij_1j_2j_3}(\omega_1;\omega_1,\omega_2,-\omega_2)\right]^k\right\}\\ &= \frac{2}{\pi\omega_1^{2\alpha}}\mathrm{P}\int_0^\infty \frac{{\omega_1'}^{2\alpha+1}\mathrm{Im}\left\{\left[\chi^{(3)}_{ij_1j_2j_3}(\omega_1';\omega_1',\omega_2,-\omega_2)\right]^k\right\}}{{\omega_1'}^2-\omega_1^2}\mathrm{d}\omega_1',\end{aligned} \tag{6.23}$$

$$\begin{aligned}&\mathrm{Im}\left\{\left[\chi^{(3)}_{ij_1j_2j_3}(\omega_1;\omega_1,\omega_2,-\omega_2)\right]^k\right\}\\ &= -\frac{2}{\pi\omega_1^{2\alpha-1}}\mathrm{P}\int_0^\infty \frac{{\omega_1'}^{2\alpha}\mathrm{Re}\left\{\left[\chi^{(3)}_{ij_1j_2j_3}(\omega_1';\omega_1',\omega_2,-\omega_2)\right]^k\right\}}{{\omega_1'}^2-\omega_1^2}\mathrm{d}\omega_1',\end{aligned} \tag{6.24}$$

where $0 \le \alpha \le 2k-1$. From the generalized K-K relations (6.23) and (6.24), using the superconvergence theorem, and considering asymptotic behavior (6.22), it is possible to derive the following set of new generalized sum rules:

$$\int_0^\infty \omega_1'^{2\alpha}\mathrm{Re}\left\{\left[\chi^{(3)}_{ij_1j_2j_3}(\omega_1';\omega_1',\omega_2,-\omega_2)\right]^k\right\}\mathrm{d}\omega_1' = 0, \quad 0 \le \alpha \le 2k-1, \tag{6.25}$$

$$\int_0^\infty \omega_1'^{2\alpha+1}\mathrm{Im}\left\{\left[\chi^{(3)}_{ij_1j_2j_3}(\omega_1';\omega_1',\omega_2,-\omega_2)\right]^k\right\}\mathrm{d}\omega_1' = 0, \quad 0 \le \alpha \le 2k-2, \tag{6.26}$$

$$\int_0^\infty \omega'^{4k-1}\mathrm{Im}\left\{\left[\chi^{(3)}_{ij_1j_2j_3}(\omega_1';\omega_1',\omega_2,-\omega_2)\right]^k\right\}\mathrm{d}\omega_1' = -\frac{\pi}{2}\left[c_{ij_1j_2j_3}(\omega_2)\right]^k. \tag{6.27}$$

These sum rules reduce to those presented in (6.18)–(6.21) if we set $k = 1$. From (6.21) and (6.27), it is possible to derive a consistency relation between the nonvanishing sum rule for the kth power of the pump-and-probe susceptibility and the kth power of the nonvanishing sum rule for conventional susceptibility:

$$-\frac{2}{\pi}\int_{0}^{\infty}\omega_1'^{4k-1}\mathrm{Im}\left\{\left[\chi^{(3)}_{ij_1j_2j_3}(\omega_1';\omega_1',\omega_2,-\omega_2)\right]^k\right\}\mathrm{d}\omega_1'$$
$$=\left[-\frac{2}{\pi}\int_{0}^{\infty}\omega_1'^3\mathrm{Im}\left\{\chi^{(3)}_{ij_1j_2j_3}(\omega_1';\omega_1',\omega_2,-\omega_2)\right\}\mathrm{d}\omega_1'\right]^k . \quad (6.28)$$

The results presented in this subsection may constitute relevant tools for the investigation of pump-and-probe experimental data. It may be easier to obtain convergence in integral properties if we consider higher powers of the susceptibility, thanks to the faster asymptotic decrease realized. A fast asymptotic decrease eases the problems related to the unavoidable spectral range finiteness.

7 Kramers-Kronig Relations and Sum Rules for Harmonic-Generation Processes

7.1 Introductory Remarks

The theoretical and experimental investigation of harmonic-generation processes is one of the most important branches of nonlinear optics [17–19]. Only recently has a complete formulation of general K-K relations and sum rules for nth-order harmonic-generation susceptibility in continuous wave approximation been obtained [31, 77, 188, 189].

A major problem in the effective verification of experimental data on the general properties of the physical quantities descriptive of harmonic-generation processes is their integral formulation, which, in principle, requires data covering the whole of the infinite positive range. Moreover, even now there are relatively few studies that report on independent measurements of the real and imaginary parts of harmonic-generation susceptibilities [28, 166] and on the validity of K-K relations in nonlinear experimental data inversion [29]. Most recently, a very detailed analysis of the integral properties of optical harmonic-generation of experimental data on polymers has been presented [32, 34].

In this chapter, we review the most relevant theoretical achievements in the determination of the general properties of harmonic-generation susceptibility. We define the nth-order harmonic-generation susceptibility tensor and obtain its analytical properties by taking advantage of the Scandolo theorem [23]. We then derive its asymptotic behavior for large values of frequency. We then combine all the information gathered, and we derive a set of independent K-K relations and sum rules for the moments of harmonic-generation susceptibility and of its powers. We also present expressions for generalized MSKK relations for the moments of harmonic-generation susceptibilities. The MSKK relations can be useful in data analysis since they relax the intrinsic limitations in the K-K approach related to the finiteness of the measured spectral range.

7.2 Application of the Scandolo Theorem to Harmonic-Generation Susceptibility

We focus on a typical experimental setup for harmonic-generation processes, where the incident radiation is given by a strictly monochromatic and linearly polarized field, so that we can express $\boldsymbol{E}(t)$ as follows:

$$E_j(t) = E_j \exp(-\mathrm{i}\omega t) + c.c. \tag{7.1}$$

Since we are interested in studying the nth-order harmonic-generation processes, we seek the $\omega_\Sigma = n\omega$ frequency component of the induced nonlinear polarization $\boldsymbol{P}^{(n)}(t)$ introduced in (5.14)–(5.16). Considering the term in (5.14), where the product of only the positive frequency components ω is considered, we obtain

$$\overline{P}_i^{(n)}(n\omega) = \chi^{(n)}_{ij_1,\dots,j_n}(n\omega;\omega,\dots,\omega)\, E_{j_1}\dots E_{j_n}. \tag{7.2}$$

We wish to emphasize that contributions to the $n\omega$ frequency component of nonlinear polarization also come from higher nonlinear orders $n+2q$, where in (5.14) we select in the argument of the susceptibility function, the positive ω frequency component $(n+q)$ times and the negative $-\omega$ frequency component q times. We will ignore these higher order contributions and concentrate on the nth-order nonlinear process.

When the frequency of the incoming radiation varies, all the arguments of harmonic-generation susceptibility in (7.2) change coherently, so that the correct straight line parametrization in $\mathbb{R}^n$ of the arguments of the susceptibility proposed in (6.4) is $\boldsymbol{v} = (1,\dots,1)$ and $\boldsymbol{w} = (0,\dots,0)$. Since all the components of the vector $\boldsymbol{v}$ are positive and hence obey condition (6.8), Scandolo's theorem [23] permits us to deduce that the harmonic-generation susceptibility is holomorphic in the upper complex ω-plane.

We also underline that the nth-order harmonic-generation susceptibility obeys the following symmetry relation, which derives from the general case presented in (5.9):

$$\chi^{(n)}_{ij_1\dots j_n}(-n\omega;-\omega,\dots,-\omega) = \left[\chi^{(n)}_{ij_1\dots j_n}(n\omega;\omega,\dots,\omega)\right]^*. \tag{7.3}$$

Such a relation implies that the real and imaginary parts are, respectively, even and odd with respect to the exchange of the sign of ω.

7.3 Asymptotic Behavior of Harmonic-Generation Susceptibility

As has been thoroughly discussed in the previous chapter in connection with the general nonlinear case, the number of independent K-K relations and sum rules that simultaneously hold for harmonic-generation susceptibility and its moments is determined by the asymptotic behavior of the susceptibility function. Therefore, the establishment of the complete general integral properties for the susceptibility considered requires microscopic treatment of the process of harmonic-generation, which we will approach using the conventional nth-order perturbation theory presented in the previous chapters, summarized by the concatenated system (2.33). Nevertheless, we have found that, while in the definition of the nth-order nonlinear polarization and susceptibility functions, the conventional length gauge for the light–matter interaction Hamiltonian presented in expression (2.24) is more efficient, the determination of the asymptotic behavior of harmonic-generation susceptibility is much more straightforward if we adopt the physically equivalent – and theoretically more elegant [38] – velocity gauge formulation presented in expression (2.21). Following the same procedure adopted in the derivation of expressions (5.21)–(5.26) in the length gauge, we find that, equivalently, nth-order nonlinear polarization $\boldsymbol{P}^{(n)}(t)$ can be expressed as follows:

$$P_i^{(n)}(t) = \int_{-\infty}^{\infty} J_{ij_1\cdots j_n}^{(n)}(t_1,\ldots,t_n)\, A_{j_1}(t-t_1)\ldots A_{j_n}(t-\tau_n)\,\mathrm{d}t_1\ldots\mathrm{d}t_n, \tag{7.4}$$

where the velocity gauge nth-order nonlinear Green function is

$$\begin{aligned} J_{ij_1\cdots j_n}^{(n)}(t_1,,\ldots,t_n) &= -\frac{e^{n+1}}{V(-\mathrm{i}\hbar mc)^n}\theta(t_1)\ldots\theta(t_n-t_{n-1}) \\ &\times \mathrm{Tr}\left\{\left[\sum_{\alpha=1}^{N} p_{j_n}^{\alpha}(-t_n),\ldots,\left[\sum_{\alpha=1}^{N} p_{j_1}^{\alpha}(-t_1),\sum_{\alpha=1}^{N} r_i^{\alpha}\right]\ldots\right]\rho(0)\right\}. \end{aligned} \tag{7.5}$$

We point out that the quadratic term of the vector potential $\boldsymbol{A}(t)$ in the interaction Hamiltonian (2.21) is not present in the definition of the Green function (7.5). The reason for this absence is that this term does not involve operators and so, being a numerical constant, its contribution is cancelled out in an iterative commutator structure as in (7.5), derived from the concatenated system of differential (2.33), with the definition (2.21) for the interaction Hamiltonian.

We observe that by applying the Fourier transform to both members of (2.22), we obtain

$$\boldsymbol{A}(\omega) = \frac{c}{\mathrm{i}\omega}\boldsymbol{E}(\omega). \tag{7.6}$$

If we compute the Fourier transform of expression (7.4), given that the incoming radiation is of the form (7.1), and take into account the result (7.6), we obtain in the velocity gauge,

$$\overline{P}_i^{(n)}(n\omega) = \Upsilon_{ij_1,\ldots,j_n}^{(n)}(n\omega;\omega,\ldots,\omega)\left(\frac{c}{\mathrm{i}\omega}\right)^n E_{j_1}\ldots E_{j_k}, \tag{7.7}$$

where

$$\Upsilon_{ij_1,\ldots,j_n}^{(n)}(n\omega;\omega,\ldots,\omega) = \int_{-\infty}^{\infty} J_{ij_1\cdots j_n}^{(n)}(\tau_1,\ldots,\tau_n)\exp\left[\mathrm{i}\omega\sum_{l=1}^{n} t_l\right]\mathrm{d}t_1\ldots\mathrm{d}t_n. \tag{7.8}$$

Since the two gauges are physically equivalent, we deduce that the following equality holds:

$$\chi^{(n)}_{ij_1,\ldots,j_n}(n\omega;\omega,\ldots,\omega) = \Upsilon^{(n)}_{ij_1,\ldots,j_n}(n\omega;\omega,\ldots,\omega)\left(\frac{c}{\mathrm{i}\omega}\right)^n. \tag{7.9}$$

Therefore, the nth-order harmonic-generation susceptibility is

$$\begin{aligned}\chi^{(n)}_{ij_1,\ldots,j_n}(n\omega,\omega,\ldots,\omega) &= -\frac{e^{n+1}}{V(\hbar m)^n\omega^n}\int_{-\infty}^{\infty}\theta(t_1)\ldots\theta(t_n-t_{n-1}) \\ &\times \mathrm{Tr}\left\{\left[\sum_{\alpha=1}^{N}p^{\alpha}_{j_n}(-t_n),\ldots,\left[\sum_{\alpha=1}^{N}p^{\alpha}_{j_1}(-t_1),\sum_{\alpha=1}^{N}r^{\alpha}_i\right]\ldots\right]\rho(0)\right\} \\ &\times \exp\left[\mathrm{i}\omega\sum_{l=1}^{n}t_l\right]\mathrm{d}t_1\ldots\mathrm{d}t_n.\end{aligned} \tag{7.10}$$

In addition, the linear susceptibility presented in expression (3.41) is included in the above expression when we set $n = 1$. In the integral in the right-hand member of expression (7.10), we apply the following variable change:

$$t_j = \sum_{i=1}^{j}\tau_i, \quad 1 \le j \le n, \tag{7.11}$$

and obtain for the general harmonic-generation susceptibility

$$\begin{aligned}\chi^{(n)}_{ij_1,\ldots,j_n}(n\omega;\omega,\ldots,\omega) &= -\frac{e^{n+1}}{V(\hbar m)^n\omega^n}\int_{-\infty}^{\infty}\theta(\tau_1)\ldots\theta(\tau_n) \\ &\times \mathrm{Tr}\left\{\left[\sum_{\alpha=1}^{N}p^{\alpha}_{j_n}(-\tau_n-\ldots-\tau_1),\ldots,\left[\sum_{\alpha=1}^{N}p^{\alpha}_{j_1}(-\tau_1),\sum_{\alpha=1}^{N}r^{\alpha}_i\right]\ldots\right]\rho(0)\right\} \\ &\times \exp\left[\mathrm{i}\omega\sum_{j=1}^{n}(n+1-j)\tau_j\right]\mathrm{d}\tau_1\ldots\mathrm{d}\tau_n.\end{aligned} \tag{7.12}$$

After a cumbersome calculation [34], we obtain the result that for large ω, the nth-order harmonic-generation susceptibility asymptotically decreases as ω^{-2n-2}:

$$\begin{aligned}\chi^{(n)}_{ij_1,\ldots,j_n}(n\omega;\omega,\ldots,\omega) =& \frac{(-1)^n}{n^2 n!}\frac{e^{n+1}}{m^{n+1}}\frac{N}{V}\mathrm{Tr}\left\{\frac{\partial^{n+1}V(\boldsymbol{r}_\alpha)}{\partial r^{\alpha}_{j_n}\ldots\partial r^{\alpha}_{j_i}\partial r^{\alpha}_i}\rho(0)\right\} \\ &\times\frac{1}{\omega^{2n+2}} + o\left(\omega^{-2n-2}\right).\end{aligned} \tag{7.13}$$

We observe that the fundamental quantum constant $\hbar$ does not appear in formula (7.13), thus suggesting that a detailed quantum physics treatment is not essential in locating this property. The quantum aspect of the expression we have obtained appears only in the definition of the expectation value of the derivatives of one-particle potential energy on the equilibrium density matrix of the system. We observe that the many-particle components of the Hamiltonian are not directly represented in this result, apart from playing a relevant role in the definition of the ground state of the system. It is possible to detect a close correspondence between the general properties of a rigorously defined quantum harmonic-generation susceptibility and those of the harmonic-generation susceptibility obtained from a simple classical anharmonic oscillator model [34, 77].

7.4 General Kramers-Kronig Relations and Sum Rules for Harmonic-Generation Susceptibility

The holomorphic properties and the asymptotic behavior of nth-order harmonic-generation susceptibility allow us to write the following set of independent K-K dispersion relations for the moments of the real and imaginary parts of the susceptibility considered for nonconducting materials [77, 189]:

$$\begin{aligned} &\omega^{2\alpha}\mathrm{Re}\left\{\chi^{(n)}_{ij_1\ldots j_n}(n\omega;\omega,\ldots,\omega)\right\} \\ &= \frac{2}{\pi}\mathrm{P}\int_0^\infty \frac{\omega'^{2\alpha+1}\mathrm{Im}\left\{\chi^{(n)}_{ij_1\ldots j_n}(n\omega';\omega',\ldots,\omega')\right\}}{\omega'^2-\omega^2}\mathrm{d}\omega', \end{aligned} \tag{7.14}$$

$$\begin{aligned} &\omega^{2\alpha-1}\mathrm{Im}\left\{\chi^{(n)}_{ij_1\ldots j_n}(n\omega;\omega,\ldots,\omega)\right\} \\ &= -\frac{2\omega}{\pi}\mathrm{P}\int_0^\infty \frac{\omega'^{2\alpha}\mathrm{Re}\left\{\chi^{(n)}_{ij_1\ldots j_n}(n\omega';\omega',\ldots,\omega')\right\}}{\omega'^2-\omega^2}\mathrm{d}\omega', \end{aligned} \tag{7.15}$$

with $0 \le \alpha \le n$, where α is such that the αth moment of the harmonic susceptibility under consideration decreases for large values of the frequency at least as fast as ω^{-2}. We observe that the number of independent K-K relations grows with the order of the process of harmonic-generation in question.

Moreover, when a function is holomorphic in a given domain and its positive integral powers share the same property, we find that the more general K-K relations stated below hold for the positive integral kth powers of the susceptibility under examination [188]:

$$\begin{aligned} &\omega^{2\alpha}\mathrm{Re}\left\{\left[\chi^{(n)}_{ij_1\ldots j_n}(n\omega;\omega,\ldots,\omega)\right]^k\right\} \\ &= \frac{2}{\pi}\mathrm{P}\int_0^\infty \frac{\omega'^{2\alpha+1}\mathrm{Im}\left\{\left[\chi^{(n)}_{ij_1\ldots j_n}(n\omega';\omega',\ldots,\omega')\right]^k\right\}}{\omega'^2-\omega^2}\mathrm{d}\omega', \end{aligned} \tag{7.16}$$

$$\omega^{2\alpha-1}\mathrm{Im}\left\{\left[\chi^{(n)}_{ij_1\dots j_n}(n\omega;\omega,\dots,\omega)\right]\right\}$$
$$= -\frac{2\omega}{\pi}\mathrm{P}\int_0^\infty \frac{\omega'^{2\alpha}\mathrm{Re}\left\{\left[\chi^{(n)}_{ij_1\dots j_n}(n\omega';\omega',\dots,\omega')\right]^k\right\}}{\omega'^2-\omega^2}\mathrm{d}\omega', \tag{7.17}$$

with $0 \le \alpha \le k(n+1)-1$. The dispersion relations presented in (7.14)–(7.17) extend the results to all orders previously obtained for the second and third orders in the past decades [170–174, 180, 190]. Comparing the asymptotic behavior obtained from expression (7.13) with those obtained by applying the superconvergence theorem [9,74] to the general K-K relations (7.16) and (7.17), we immediately obtain the following set of sum rules:

$$\int_0^\infty \omega'^{2\alpha}\mathrm{Re}\left\{\left[\chi^{(n)}_{ij_1\dots j_n}(n\omega',\omega',\dots,\omega')\right]^k\right\}\mathrm{d}\omega' = 0, \quad 0 \le \alpha \le k(n+1)-1, \tag{7.18}$$

$$\int_0^\infty \omega'^{2\alpha+1}\mathrm{Im}\left\{\left[\chi^{(n)}_{ij_1\dots j_n}(n\omega',\omega',\dots,\omega')\right]^k\right\}\mathrm{d}\omega' = 0, \quad 0 \le \alpha \le k(n+1)-2, \tag{7.19}$$

$$\int_0^\infty \omega'^{2k(n+1)-1}\mathrm{Im}\left\{\left[\chi^{(n)}_{ij_1\dots j_n}(n\omega',\omega',\dots,\omega')\right]^k\right\}\mathrm{d}\omega'$$
$$= \frac{\pi}{2}\left[\frac{(-1)^{n+1}}{n^2 n!}\frac{e^{n+1}}{m^{n+1}}\frac{N}{V}\mathrm{Tr}\left\{\frac{\partial^{n+1}V(\boldsymbol{r}_\alpha)}{\partial r^\alpha_{j_n}\dots\partial r^\alpha_{j_i}\partial r^\alpha_i}\rho(0)\right\}\right]^k. \tag{7.20}$$

All the moments of the kth power of nth-order harmonic-generation susceptibility vanish except that of order $2k(n+1)-1$ of the imaginary part. In the most fundamental case of $k=1$, the nonvanishing sum rule creates a conceptual bridge between a measure of the $(n+1)$th-order nonlinearity of the potential energy of the system and the measurements of the imaginary part of the susceptibility under examination throughout the spectrum, thus relating structural and optical properties of the material under consideration, i.e. the *static* and *dynamic* properties of the electronic system.

We can also observe that from the nonvanishing sum rule in (7.20), it is possible to obtain a simple consistency relation between the nonvanishing sum rule for the kth power of harmonic-generation susceptibility and the kth power of the nonvanishing sum rule for conventional susceptibility, where we select $k=1$:

$$-\frac{2}{\pi}\int_0^\infty \omega'^{2k(n+1)-1}\mathrm{Im}\left\{\left[\chi^{(n)}_{ij_1\dots j_n}(n\omega',\omega',\dots,\omega')\right]^k\right\}\mathrm{d}\omega'$$
$$= \left[-\frac{2}{\pi}\int_0^\infty \omega'^{2n+1}\mathrm{Im}\left\{\chi^{(n)}_{ij_1\dots j_n}(n\omega',\omega',\dots,\omega')\right\}\mathrm{d}\omega'\right]^k. \tag{7.21}$$

The constraints presented here are, in principle, universal, since they essentially derive from the principle of causality in the response of the matter to external radiation, and so they are expected to hold for any material. They provide fundamental tests of self-consistency that any experimental or model generated data have to obey. We point out that verification of K-K relations and sum rules constitutes an unavoidable benchmark for any investigation into the nonlinear response of matter to radiation over a wide spectral range.

We wish to underline that, similarly to the linear case, the integral properties obtained by adopting a full ab initio quantum mechanical approach show very close correspondence with the results obtained in [77] with a simple nonlinear oscillator model, provided we consider the expectation value of the derivatives of the potential energy as the quantum analogue of the same derivatives of the classical potential energy evaluated at the equilibrium position. The main reason for this correspondence is that the basic ingredients of the general integral relations are the analytical properties and asymptotic behavior of harmonic-generation susceptibility. These properties do not depend on the microscopic treatment of the interaction between light and matter but are connected to the validity of the causality principle in physical systems [2,3,34,35,38]. Similar results have been obtained for the susceptibility function relevant to pump-and-probe processes [34,129].

7.4.1 General Integral Properties of Nonlinear Conductors

Conducting materials are characterized by the presence of nonvanishing static conductance, which changes the integral properties of their nth-order harmonic-generation susceptibilities, as in linear optics [68–70,78]. In order to avoid an excessively cumbersome study, we will not present the K-K relations for the higher powers of the nth-order harmonic susceptibility of metals.

Remembering that susceptibility can be expressed in terms of conductivity in every order of nonlinearity,

$$\chi^{(n)}_{ij_1\cdots j_n}(n\omega;\omega,\ldots,\omega) = \mathrm{i}\frac{\sigma^{(n)}_{ij_1\cdots j_n}(n\omega;\omega,\ldots,\omega)}{n\omega}, \tag{7.22}$$

in the case of conductors, by definition, we can express harmonic-generation susceptibility in terms of the nonvanishing real tensor of nonlinear static conductance of order n for frequencies close to zero:

$$\chi^{(n)}_{ij_1\cdots j_n}(n\omega;\omega,\ldots,\omega)\Big|_{\omega\approx 0} \approx \mathrm{i}\frac{\sigma^{(n)}_{ij_1\cdots j_n}(0)}{n\omega}. \tag{7.23}$$

The presence of this pole at the origin of the ω-axis changes the K-K relation (7.15) for $\alpha = 0$:

$$
\begin{aligned}
&\operatorname{Im}\left\{\chi^{(n)}_{ij_1\ldots j_n}(n\omega;\omega,\ldots,\omega)\right\} - \frac{\sigma^{(n)}_{ij_1\ldots j_n}(0)}{n\omega} \\
&= -\frac{2\omega}{\pi}\mathrm{P}\int_0^\infty \frac{\operatorname{Re}\left\{\chi^{(n)}_{ij_1\ldots j_n}(n\omega';\omega',\ldots,\omega')\right\}}{\omega'^2-\omega^2}\mathrm{d}\omega' .
\end{aligned}
\tag{7.24}
$$

By applying the superconvergence theorem to the dispersion relation (7.24), we derive the related sum rule:

$$
\int_0^\infty \operatorname{Re}\left\{\chi^{(n)}_{ij_1\ldots j_n}(n\omega';\omega',\ldots,\omega')\right\}\mathrm{d}\omega' = -\frac{\pi}{2n}\sigma^{(n)}_{ij_1\ldots j_n}(0)\,. \tag{7.25}
$$

Equations (7.24) and (7.25) extend the results for linear optics to the nonlinear case, previously presented in Sect. 4.3.1 and in (4.19), which have been carefully verified with experimental data [68–70, 78].

All the K-K relations (7.14) and (7.15) and related sum rules (7.18)–(7.20) that we obtain by setting $1 \le \alpha \le n$ remain unchanged for conductors because the higher moments of harmonic-generation susceptibility do not have a pole at the origin.

We propose that the established general integral properties for harmonic-generation processes on conductors should be experimentally verified.

7.5 Subtractive Kramers-Kronig Relations for Harmonic-Generation Susceptibility

The characteristic integral structure of K-K relations requires knowledge of the spectrum in a semi-infinite angular frequency range. Unfortunately, in practical spectroscopy, only a finite spectral range can be measured. Moreover, technical difficulties in gathering information about nonlinear optical properties over a sufficiently wide spectral range make even the application of approximate K-K relations problematic. The extrapolations in K-K analysis, such as estimation of data beyond the measured spectral range, can be a serious source of errors [13, 71]. Recently, King [191] presented an efficient numerical approach to the evaluation of K-K relations. Nevertheless, the problem of data fitting is always present in regions outside the measured range.

In the context of linear optics, SSKK [109] and MSKK [110] relations have been proposed in order to relax the limitations caused by finite-range data, as described in Sect. 4.7.2.

Subtractive K-K relations have been proposed only recently in the context of nonlinear optics and especially for harmonic-generation susceptibility [31]. The idea behind the subtractive Kramers-Kronig technique is that inversion of the real (imaginary) part of nth-order harmonic-generation susceptibility can be greatly improved if we have one or more anchor points, i.e. a single or multiple measurement of the imaginary (real) part for a set of frequencies. For simplicity, in this section, we use the following simplified notation:

$$\chi^{(n)}_{ij_1\ldots j_n}(n\omega;\omega,\ldots,\omega)=\chi^{(n)}_{ij_1\ldots j_n}(n\omega). \tag{7.26}$$

With the aid of mathematical induction (see Appendix A in [110]), we can derive the multiply subtractive K-K relations for real and imaginary parts as follows:

$$\begin{aligned}
&\omega^{2\alpha}\mathrm{Re}\left\{\chi^{(n)}_{ij_1\ldots j_n}(n\omega)\right\}\\
&=\left[\frac{(\omega^2-\omega_2^2)(\omega^2-\omega_3^2)\cdots(\omega^2-\omega_Q^2)}{(\omega_1^2-\omega_2^2)(\omega_1^2-\omega_3^2)\cdots(\omega_1^2-\omega_Q^2)}\right]\omega_1^{2\alpha}\mathrm{Re}\left\{\chi^{(n)}_{ij_1\ldots j_n}(n\omega_1)\right\}+\cdots\\
&\quad+\left[\frac{(\omega^2-\omega_1^2)\cdots(\omega^2-\omega_{j-1}^2)(\omega^2-\omega_{j+1}^2)\cdots(\omega^2-\omega_Q^2)}{(\omega_j^2-\omega_1^2)\cdots(\omega_j^2-\omega_{j-1}^2)(\omega_j^2-\omega_{j+1}^2)\cdots(\omega_j^2-\omega_Q^2)}\right]\\
&\quad\times\omega_j^{2\alpha_j}\mathrm{Re}\left\{\chi^{(n)}_{ij_1\ldots j_n}(n\omega_j)\right\}+\cdots\\
&\quad+\left[\frac{(\omega^2-\omega_1^2)(\omega^2-\omega_2^2)\cdots(\omega^2-\omega_{Q-1}^2)}{(\omega_Q^2-\omega_1^2)(\omega_Q^2-\omega_2^2)\cdots(\omega_Q^2-\omega_{Q-1}^2)}\right]\omega_Q^{2\alpha}\mathrm{Re}\left\{\chi^{(n)}_{ij_1\ldots j_n}(n\omega_Q)\right\}+\\
&\quad+\frac{2}{\pi}\left[(\omega^2-\omega_1^2)(\omega^2-\omega_2^2)\cdots(\omega^2-\omega_Q^2)\right]\\
&\quad\times\mathrm{P}\int_0^\infty\frac{\omega'^{2\alpha+1}\mathrm{Im}\left\{\chi^{(n)}_{ij_1\ldots j_n}(n\omega')\right\}}{(\omega'^2-\omega^2)(\omega'^2-\omega_1^2)\cdots(\omega'^2-\omega_Q^2)}\mathrm{d}\omega',
\end{aligned}\tag{7.27}$$

$$\begin{aligned}
&\omega^{2\alpha-1}\mathrm{Im}\left\{\chi^{(n)}_{ij_1\ldots j_n}(n\omega)\right\}\\
&=\left[\frac{(\omega^2-\omega_2^2)(\omega^2-\omega_3^2)\cdots(\omega^2-\omega_Q^2)}{(\omega_1^2-\omega_2^2)(\omega_1^2-\omega_3^2)\cdots(\omega_1^2-\omega_Q^2)}\right]\omega_1^{2\alpha-1}\mathrm{Im}\left\{\chi^{(n)}_{ij_1\ldots j_n}(n\omega_1)\right\}+\cdots\\
&\quad+\left[\frac{(\omega^2-\omega_1^2)\cdots(\omega^2-\omega_{j-1}^2)(\omega^2-\omega_{j+1}^2)\cdots(\omega^2-\omega_Q^2)}{(\omega_j^2-\omega_1^2)\cdots(\omega_j^2-\omega_{j-1}^2)(\omega_j^2-\omega_{j+1}^2)\cdots(\omega_j^2-\omega_Q^2)}\right]\\
&\quad\times\omega_j^{2\alpha-1}\mathrm{Im}\left\{\chi^{(n)}_{ij_1\ldots j_n}(n\omega_j)\right\}+\cdots\\
&\quad+\left[\frac{(\omega^2-\omega_1^2)(\omega^2-\omega_2^2)\cdots(\omega^2-\omega_{Q-1}^2)}{(\omega_Q^2-\omega_1^2)(\omega_Q^2-\omega_2^2)\cdots(\omega_Q^2-\omega_{Q-1}^2)}\right]\omega_Q^{2\alpha-1}\mathrm{Im}\left\{\chi^{(n)}_{ij_1\ldots j_n}(n\omega_Q)\right\}+\\
&\quad-\frac{2}{\pi}\left[(\omega^2-\omega_1^2)(\omega^2-\omega_2^2)\cdots(\omega^2-\omega_Q^2)\right]\\
&\quad\times\mathrm{P}\int_0^\infty\frac{\omega'^{2\alpha}\mathrm{Re}\left\{\chi^{(n)}_{ij_1\ldots j_n}(n\omega')\right\}}{(\omega'^2-\omega^2)(\omega'^2-\omega_1^2)\cdots(\omega'^2-\omega_Q^2)}\mathrm{d}\omega',
\end{aligned}\tag{7.28}$$

with $0 \leq \alpha \leq n$; here ω_j with $j = 1, \cdots, Q$ denote the anchor points. Note that the anchor points in (7.27) and (7.28) need not be the same. We observe that the integrands of Eqs. (7.27) and (7.28) show a remarkably faster asymptotic decrease as a function of angular frequency, than the conventional K-K relations given by (7.14) and (7.15). In the case of Q-times subtracted K-K relations, the integrands asymptotically decrease as $\omega^{2\alpha-(2n+2+2Q)}$, whereas in the case of conventional K-K relations, the decrease is proportional to $\omega^{2\alpha-(2n+2)}$. Hence, it can be expected that the limitations related to the presence of an experimentally unavoidable finite frequency range are relaxed, and the precision of the integral inversions is then enhanced.

The additional information for optical data inversion derived from the knowledge of anchor points is useful mostly when they are located well outside the main spectral features of the range considered. The first reason is that if we already have direct access to information on the main features of the spectrum, optical data inversion may be unnecessary. Moreover, even small relative errors in the experimental value of anchor points located in the main features of the spectrum would *propagate* in (7.27) or (7.28) over all the spectral region under examination so that the relative error in the out-of-resonance portions would be considerable, thus causing large error in the estimation of the complex phase. We note that according to Palmer et al. [110], under simplifying assumptions for the spectral shape, the choice of the anchor points is optimal when they are near the zeros of the Qth-order Chebyshev polynomial of the first kind.

In linear optical spectroscopy it is usually easy to get information on optical constants at various anchor points. However, in the field of nonlinear optics, it may be difficult to obtain the real and imaginary parts of nonlinear susceptibility at various anchor points. We emphasize that even a single anchor point reduces the errors caused by a finite spectral range in data inversion of nonlinear optical data, as shown in the next chapter. For one anchor point, say, at frequency ω_1, we obtain the following singly subtractive K-K relations from (7.27) and (7.28):

$$\begin{aligned}\omega^{2\alpha}\mathrm{Re}\left\{\chi^{(n)}_{ij_1\cdots j_n}(n\omega)\right\} &= \omega_1^{2\alpha}\mathrm{Re}\left\{\chi^{(n)}_{ij_1\cdots j_n}(n\omega_1)\right\} \\ &+ \frac{2(\omega^2-\omega_1^2)}{\pi}\mathrm{P}\int_0^\infty \frac{\omega'^{2\alpha+1}\mathrm{Im}\left\{\chi^{(n)}_{ij_1\cdots j_n}(n\omega')\right\}}{(\omega'^2-\omega^2)(\omega'^2-\omega_1^2)}\mathrm{d}\omega',\end{aligned} \tag{7.29}$$

$$\begin{aligned}\omega^{2\alpha-1}\mathrm{Im}\left\{\chi^{(n)}_{ij_1\cdots j_n}(n\omega)\right\} &= \omega_1^{2\alpha-1}\mathrm{Im}\left\{\chi^{(n)}(n\omega_1)\right\} \\ &- \frac{2(\omega^2-\omega_1^2)}{\pi}\mathrm{P}\int_0^\infty \frac{\omega'^{2\alpha}\mathrm{Re}\left\{\chi^{(n)}_{ij_1\cdots j_n}(n\omega')\right\}}{(\omega'^2-\omega^2)(\omega'^2-\omega_1^2)}\mathrm{d}\omega',\end{aligned} \tag{7.30}$$

where $0 \leq \alpha \leq n$.

8 Kramers-Kronig Relations and Sum Rules for Data Analysis: Examples

8.1 Introductory Remarks

Only in a few investigations have independent measurements of the real and imaginary parts of harmonic-generation susceptibilities been performed across a relatively wide range [28, 29, 143, 166]. Consequently, verification of the coherence of measured data by checking the self-consistency of Kramers-Kronig relations is still of very limited use [29]. Until very recently, no studies at all have dealt with the experimental verification of sum rules.

In this chapter, we report the major results of the first analyses of harmonic-generation data where the full potential of the generalized K-K relations and sum rules for harmonic-generation susceptibilities as well as of the subtractive K-K relations presented in the previous chapter is exploited [31, 32, 34].

8.2 Applications of Kramers-Kronig Relations for Data Inversion

The calculations presented here are based on two published sets of experimental data on third-harmonic-generation in polymers, where the real and imaginary parts of susceptibility were independently measured. The first data set refers to measurements taken on polisylane [29] and spans a frequency range of 0.4–2.5 eV. The second data set refers to measurements taken on polythiophene [166] and spans a frequency range of 0.5–2.0 eV.

We consider the worst scenario, namely, data from a limited spectral range and with no knowledge of anchor points [31, 109, 110]. We do not assume any asymptotic behavior outside the data range but use only the experimental data, since extrapolation is somewhat arbitrary and in K-K analysis can be quite problematic [13, 15]. There is no information about the tensorial components of harmonic-generation susceptibility that have been measured, so that we adopt a simplified scalar notation. We also denote $\chi^{(3)}(3\omega;\omega,\omega,\omega)$ by $\chi^{(3)}(3\omega)$ to simplify the notation.

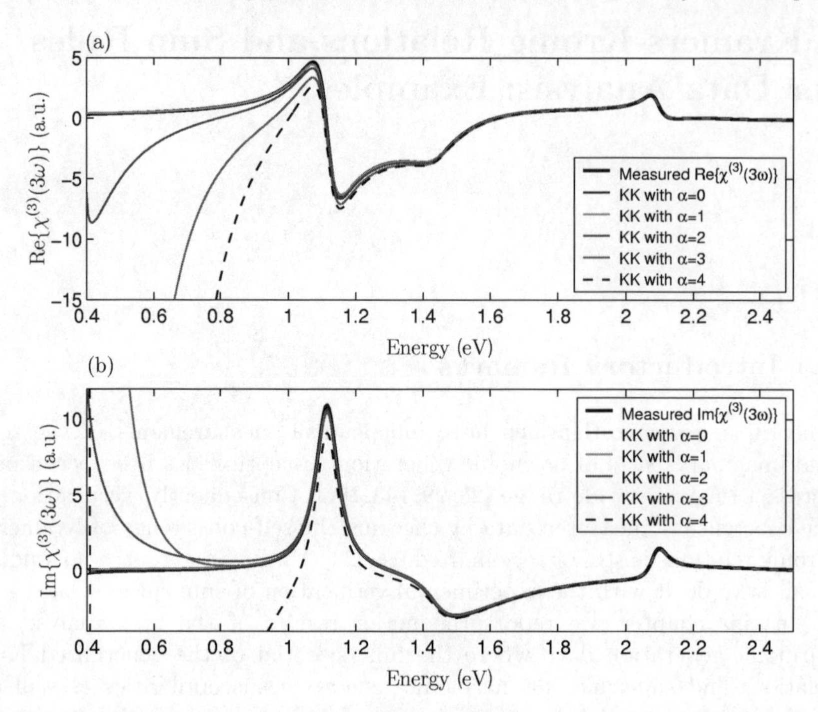

Fig. 8.1. Efficacy of K-K relations in retrieving (**a**) $\omega^{2\alpha}\mathrm{Re}\{\chi^{(3)}(3\omega)\}$ and (**b**) $\omega^{2\alpha+1}\mathrm{Im}\{\chi^{(3)}(3\omega)\}$ on polysilane. Reproduced from [32]

8.2.1 Kramers-Kronig Inversion of Harmonic-Generation Susceptibility

The self-consistency of the red and imagenary parts of the two data sets can be checked by observing the efficacy of the K-K relations in inverting the optical data for the functions $\omega^{2\alpha}[\chi^{(3)}(3\omega)]^k$, with $k = 1, 2$. We report the results obtained by applying truncated K-K relations and use a self-consistent procedure [32, 34].

In the paper by Kishida et al. [29], a check of the validity of the K-K relations was performed by comparing the measured and retrieved $\chi^{(3)}(3\omega)$. Consideration of the moments of susceptibility is not a mere add-on to the work by Kishida et al. [29]; it represents a fundamental conceptual improvement. These additional independent relations are peculiar to nonlinear phenomena and provide independent double-checks of the experimental data that must be obeyed in addition to conventional K-K relations. In Fig. 8.1a,b, we present the results of the K-K inversion for, respectively, the real and imaginary parts of the third-harmonic-generation susceptibility data on polysilane. We observe that in both cases the retrieved data obtained with the choices $\alpha = 0, 1$ are almost indistinguishable from the experimental data, while for $\alpha = 2$ and $\alpha = 3$, the agreement is quite poor in the lower part of the spec-

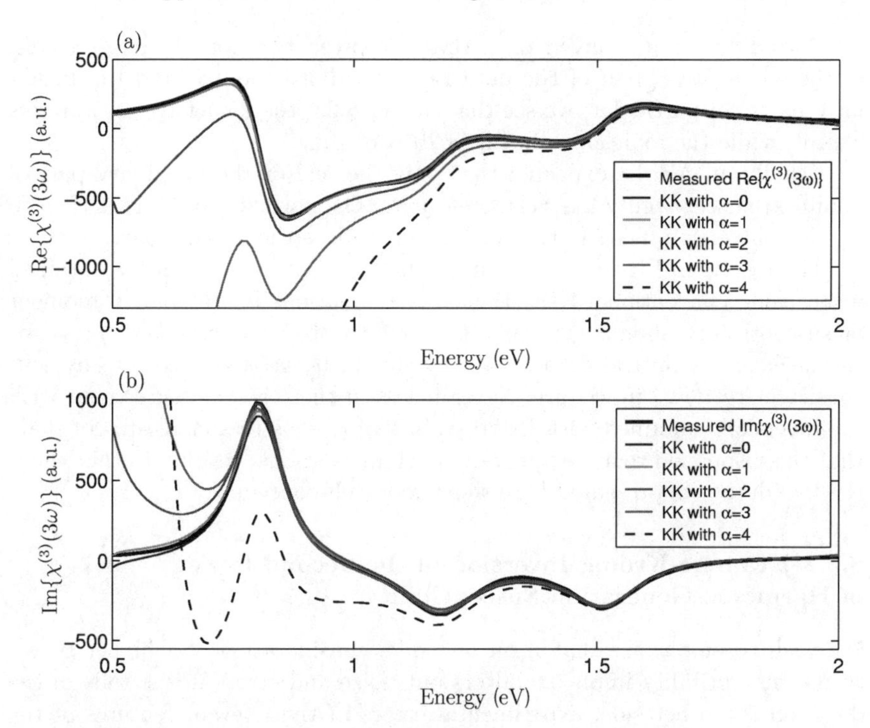

Fig. 8.2. Efficacy of K-K relations in retrieving (**a**) $\omega^{2\alpha}\mathrm{Re}\{\chi^{(3)}(3\omega)\}$ and (**b**) $\omega^{2\alpha+1}\mathrm{Im}\{\chi^{(3)}(3\omega)\}$ on polytiophene. Reproduced from [32]

trum. The error induced by the presence of the cutoff in the high-frequency range becomes more critical in the data inversion for larger α, since a slower asymptotic decrease is realized. It is reasonable to expect that by inverting the data with the additional information given by anchor points located in the lower part of the data range, these divergences can be cured. The theory suggests that for $\alpha = 4$, no convergence should occur. Actually, we observe that, while the main features around 1.1 eV are represented, there is no convergence at all for the lower frequencies; the absence of a clear transition in retrieving performance between the $\alpha = 3$ and $\alpha = 4$ cases is due to the finiteness of the data range.

In Figs. 8.2a and 8.2b, we show a comparison between the retrieved and experimental data of third-harmonic-generation susceptibility on polythiophene for real and imaginary parts, respectively. The dependence of the accuracy of the quality of the data inversion is similar to the previous case: for $\alpha = 0, 1$, the agreement is virtually perfect, while for $\alpha = 2, 3$, we have progressively worse performance in the low-frequency range. Nonetheless, the peaks in the imaginary part are still well reproduced, while the dispersive structures in the real part are present but shifted toward lower values. In this case, the quality of the retrieved data for $\alpha = 4$ and $\alpha = 3$ is more distinct than in the previous data set. The inversion with $\alpha = 4$ presents a notable disagreement in the whole lower half of the data range for both the real and the imaginary parts. In particular, we see that in Fig. 8.2a, the dispersive structure is absent, while the main peak in Fig. 8.2b is missing.

Usually, it may be expected that only the real or the imaginary part of nonlinear susceptibility has been measured. The normal procedure is then to try data inversion using K-K in order to calculate the missing part.

The results in Figs. 8.1–8.2 confirm that the best convergence is obtained when using conventional K-K. Hence, K-K relations for the $\alpha = 0$ moment of susceptibility should generally be used to obtain a *first best guess* for the inversion of optical data, and they should be used as seed for any self-consistent retrieval procedure. Nevertheless, if there is good agreement with the inversions obtained with higher values of α, it is reasonable to conclude that the dispersion relations provide much more robust results. In this sense, the two data sets presented here show good self-consistency.

8.2.2 Kramers-Kronig Inversion of the Second Power of Harmonic-Generation Susceptibility

We wish to emphasize that if on one side, consideration of a higher power of the susceptibility implicitly filters out noise and errors in the tails of the data, on the other side, experimental errors in the relevant features of the spectrum – peaks for the imaginary part and dispersive structures for the real part – are greatly enhanced if the higher powers of the susceptibility are considered. In the latter case, consistency between the K-K inversion of the different moments is expected to be more problematic than in the $k = 1$ case. Therefore, improved convergence for more moments will occur for the powers of susceptibility $k > 1$, *only if* the data are basically good [32, 34].

In Fig. 8.3a,b, we show the results of K-K inversion for the real and imaginary parts of the second power of third-harmonic-generation susceptibility on polysilane. We observe that for $\alpha = 0, 1, 2$, the agreement between the experimental and retrieved data is almost perfect, while it becomes progressively worse for increasing α. Nevertheless, as long as $\alpha \leq 6$, the main features are reproduced well for both the real and the imaginary parts, and the retrieved data match well if the photon energy is $\geq 1.0\,\mathrm{eV}$.

In Fig. 8.4a,b, we show the result of applying K-K relations to the second power of the susceptibility data taken on polythiophene. In this case, the agreement is also very good if $\alpha = 0, 1, 2$, but the narrower frequency range does not permit data inversion if very high moments are considered. If we consider the real part – Fig. 8.4a – for $\alpha = 3, 4$, the K-K data inversion provides good reproduction of the experimental data also for photon energies $\leq 0.7\,\mathrm{eV}$. For $\alpha \geq 5$, there is no convergence in the lower half of the spectral range. For the imaginary part – Fig. 8.4b – we can repeat the same observations, except that for $\alpha = 5$, there still is good reproduction of the main features of the curve.

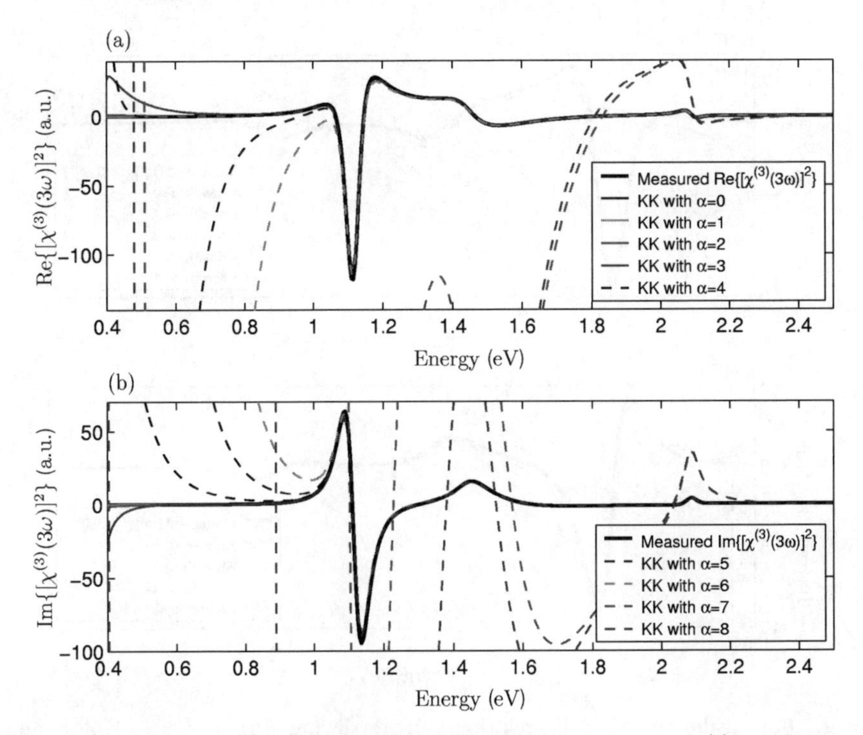

Fig. 8.3. Efficacy of K-K relations in retrieving (**a**) $\mathrm{Re}\{[\chi^{(3)}(3\omega)]^2\}$ and (**b**) $\mathrm{Im}\{[\chi^{(3)}(3\omega)]^2\}$ on polysilane. The line legends for the results of the K-K relations are common to both figures. Reproduced from [32]

The theory predicts convergence for K-K relations with $\alpha \leq 7$ and divergence for $\alpha = 8$. In our analysis, we have divergence already for $\alpha = 7$ in the case of polysilane and for $\alpha = 6, 7$ in the case of polythiophene data. The disagreement between the theory and the experiment can be safely attributed to the truncation occurring in the high-frequency range. The poor representation of far asymptotic behavior affects mostly the convergence of dispersion relations of very high moments.

We emphasize that if only one of the real or imaginary parts of the harmonic-generation susceptibility has been measured experimentally, there is no direct use of K-K relations relative to the higher powers of the susceptibility, since the multiplication mixes the real and imaginary parts. Hence, in this case, the K-K relations for $k > 1$ can be used as tests for the robustness of the results obtained with the dispersion relations applied to conventional susceptibility.

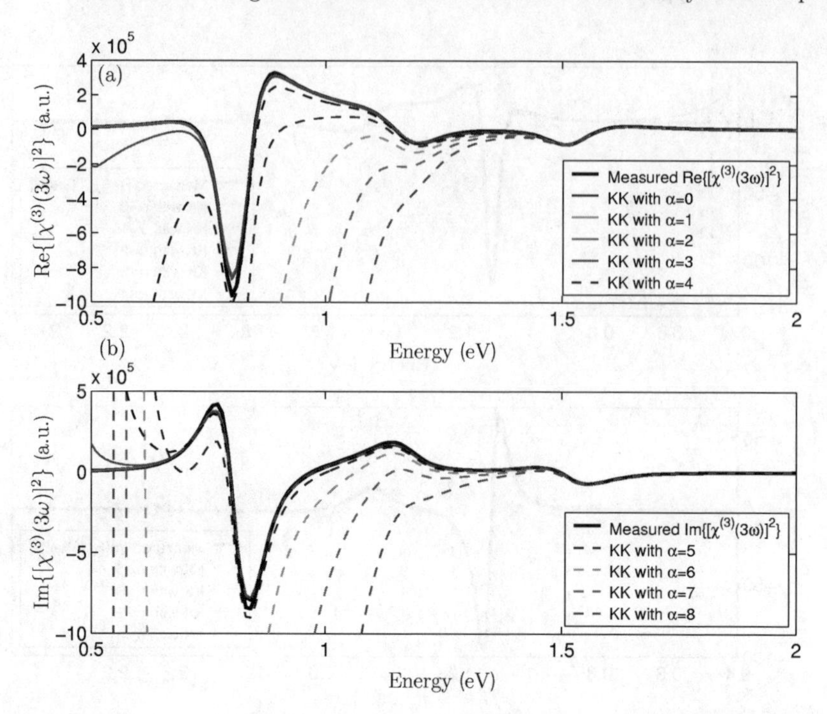

Fig. 8.4. Efficacy of K-K relations in retrieving (**a**) $\mathrm{Re}\{[\chi^{(3)}(3\omega)]^2\}$ and (**b**) $\mathrm{Im}\{[\chi^{(3)}(3\omega)]^2\}$ on polythiophene. The line legends for the results of the K-K relations are common to both figures. Reproduced from [32]

8.3 Verification of Sum Rules for Harmonic-Generation Susceptibility

In general, good accuracy in the verification of sum rules is more difficult to achieve than for K-K relations. In consequence, a positive outcome of this test provides a very strong argument in support of the quality and coherence of the experimental data [75,76].

In the case of harmonic nonlinear processes, the technical constraints for acquiring information about a very wide frequency range are very severe, and verification of the sum rules is critical, especially for those involving relatively large values of α which determine a slower asymptotic decrease. Nevertheless, if we consider increasingly large values of k, the integrands in (7.18) and (7.19) have a much faster asymptotic decrease, so that the missing high-frequency tails tend to become negligible. Therefore, we expect that for a given α, the convergence of the sum rules should be more accurate for higher values of k, if we assume that the main features of the spectrum are well reproduced by the experimental data, as explained in the previous section.

We focus first on the vanishing sum rules (7.18)–(7.19). In order to have a measure of how precisely the vanishing sum rules are obeyed for the two

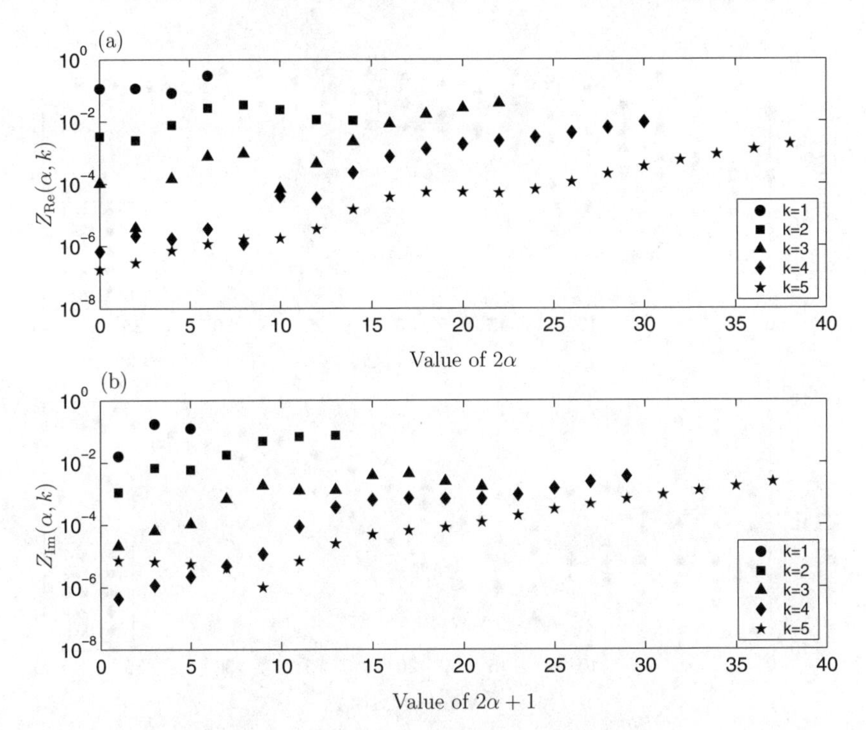

Fig. 8.5. Convergence to zero of the vanishing sum rules (**a**) $\omega^{2\alpha}\mathrm{Re}\{[\chi^{(3)}(3\omega)]^k\}$ and (**b**) $\omega^{2\alpha+1}\mathrm{Im}\{[\chi^{(3)}(3\omega)]^k\}$ with $1 \le k \le 5$, data on polysilane. Reproduced from [32]

experimental data sets under examination, along the lines of Altarelli et al. [9,78], we introduce the dimensionless quantities Z_{Re} and Z_{Im}:

$$Z_{\mathrm{Re}}(\alpha,k) = \left| \frac{\int_{\omega_{\min}}^{\omega_{\max}} \omega^{2\alpha}\mathrm{Re}\{[\chi^{(3)}(3\omega)]^k\}\mathrm{d}\omega}{\int_{\omega_{\min}}^{\omega_{\max}} \omega^{2\alpha}|\mathrm{Re}\{[\chi^{(3)}(3\omega)]^k\}|\mathrm{d}\omega} \right|, \tag{8.1}$$

$$Z_{\mathrm{Im}}(\alpha,k) = \left| \frac{\int_{\omega_{\min}}^{\omega_{\max}} \omega^{2\alpha+1}\mathrm{Im}\{[\chi^{(3)}(3\omega)]^k\}\mathrm{d}\omega}{\int_{\omega_{\min}}^{\omega_{\max}} \omega^{2\alpha+1}|\mathrm{Im}\{[\chi^{(3)}(3\omega)]^k\}|\mathrm{d}\omega} \right|. \tag{8.2}$$

Low values of $Z_{\mathrm{Re}}(\alpha,k)$ and $Z_{\mathrm{Im}}(\alpha,k)$ imply that the negative and positive contributions to the corresponding sum rule cancel out quite precisely compared to their total absolute values. The two data sets for the polymers have quite different performance in the verification of these sum rules.

In Fig. 8.5a,b, we present the results obtained with the data taken on polysilane by considering $1 \le k \le 5$ for, respectively, the sum rules of the real and the imaginary parts. We can draw very similar conclusions in both cases. Generally, we see that for a given α, we have better convergence when a higher k is considered, with a remarkable increase in the accuracy of the sum rules for

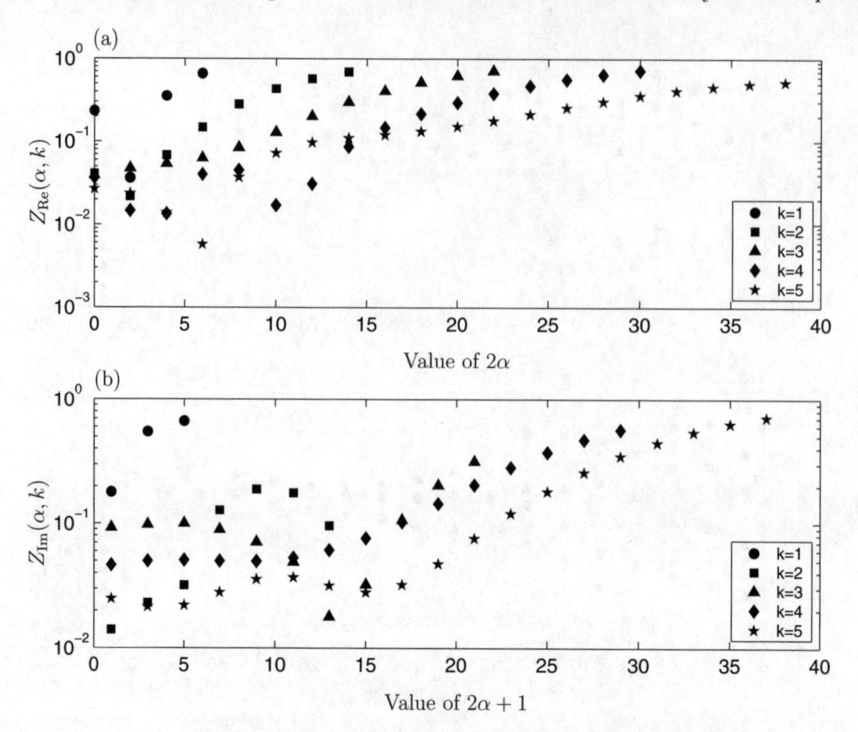

Fig. 8.6. Convergence to zero of the vanishing sum rules (**a**) $\omega^{2\alpha}\mathrm{Re}\{[\chi^{(3)}(3\omega)]^k\}$ and (**b**) $\omega^{2\alpha+1}\mathrm{Im}\{[\chi^{(3)}(3\omega)]^k\}$ with $1 \leq k \leq 5$, data on polythiophene. Reproduced from [32]

$k \geq 3$. Consistent with the argument that the speed of asymptotic behavior is critical in determining the accuracy of the sum rule, we generally also have a decrease in the quality of the convergence to zero when, for a given k, higher moments are considered, thus increasing the value of α. Particularly impressive is the increase in the performance in the convergence of the sum rules of $\chi^{(3)}(3\omega)$ for both the real part$(2\alpha = 0, 2, 4, 6)$ and the imaginary part $(2\alpha + 1 = 1, 3, 5)$ when we consider $k = 4, 5$ instead of $k = 1$. The values of Z_{Re} and Z_{Im} decrease by more than three orders of magnitude in all cases considered.

In Fig. 8.6a,b, we present the corresponding results for the experimental data found for polythiophene. Most of the sum rules computed with this data set show very poor convergence to zero, since the corresponding Z_{Im} and Z_{Re} are above 10^{-1}. Nevertheless, we can draw conclusions similar to those in the previous case in terms of change in the accuracy of the convergence for different values of k and α. Consistent with the relevance of the asymptotic behavior, the precision increases with increasing k and with decreasing α. But in this case, for a given α, the improvement in the convergence of the sum rules obtained by considering a high value of k, instead of $k = 1$, is generally small, in most cases consisting of the decrease of Z_{Im} and Z_{Re} below or around one order of magnitude.

The two data sets for polymers differ greatly in the precision achieved in the verification of the sum rules. In many corresponding cases, the polysilane data provide results that are better by orders of magnitude. The main reason for this discrepancy is the much stronger dependence of the sum rule precision on the position of the high-frequency range experimental cutoff relative to the saturation of the electronic transitions of the material. It is likely that the data on polythiophene, apart from being narrower in absolute terms, provide a more limited description of the main electronic properties of the material. This result is consistent with the previously presented slightly worse performance of this data set in the K-K inversion of the second power of $\chi^{(3)}(3\omega)$, where the relevance of the out-of-range data is also quite prominent.

For both of the two data sets considered in this study, there is no consistency between the nonvanishing sum rules referring to the various powers $1 \leq k \leq 5$ of the susceptibility under examination, since the previously established consistency relation (7.21) is essentially not obeyed. Hence, the structural constant determined by the value resulting from the integration of the highest moment of the imaginary part of the harmonic-generation susceptibility presented in (7.20) cannot be reliably evaluated with the data sets we are considering. The poor performance of the experimental data in reproducing this theoretical result is essentially due to the fact that we are dealing with the slowest converging sum rules for each k. These depend delicately on the asymptotic behavior of the data, which, as already noted, is relatively poorly represented, given the narrowness of the data under analysis. We conclude that better data covering a much wider spectral range are required to deal effectively with nonvanishing sum rules.

8.4 Application of Singly Subtractive Kramers-Kronig Relations

Here we apply singly subtractive K-K relations (7.29)–(7.30) for the analysis of the experimental values of the real and imaginary parts of the nonlinear susceptibility of the previously presented third-order harmonic wave generation on polysilane, obtained by Kishida et al. [29]. We will henceforth use a scalar notation because we have no information on the tensorial components involved in the measurements.

First, we consider only data ranging from 0.9 to 1.4 eV in order to simulate low data availability, and then we compare the quality of the data inversion obtained with the $\alpha = 0$ conventional K-K and SSKK relations within this energy range [31]. The interval 0.9–1.4 eV constitutes a good test since it contains the most relevant features of both parts of the susceptibility. However, a lot of the spectral structure is left outside the interval and the asymptotic

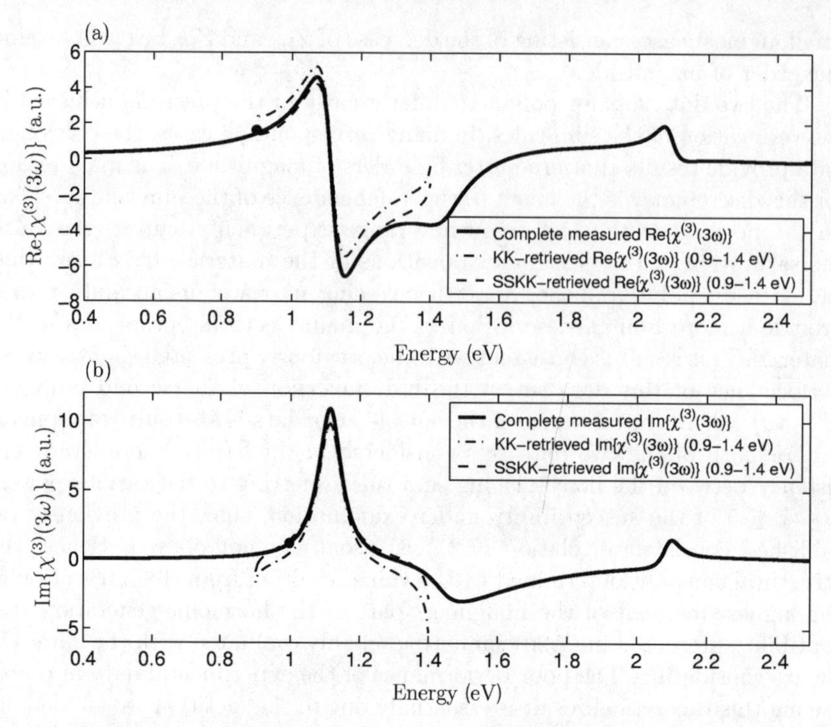

Fig. 8.7. Efficacy of SSKK vs. K-K relations in retrieving (**a**) $\mathrm{Re}\{\chi^{(3)}(3\omega)\}$ and (**b**) $\mathrm{Im}\{\chi^{(3)}(3\omega)\}$ on polysilane (•=anchor point). Reproduced from [31]

behavior is not established for either part. Hence, no plain optimal conditions for optical data inversion are established.

In Fig. 8.7a, we show the results obtained for the real part of the third-order harmonic-generation susceptibility. The solid line in Fig. 8.7a represents the experimental data. The dash-dotted line in Fig. 8.7a, which was calculated by using a conventional K-K relation by truncated integration of (7.14), consistently gives a poor match with the actual line. In contrast, we obtain better agreement with a single anchor point located at $\omega_1 = 0.9\,\mathrm{eV}$, which is represented by the dashed line in Fig. 8.7a. SSKK and measured data for the real part of the susceptibility are almost indistinguishable up to 1.3 eV.

In Fig. 8.7b, calculations similar to those presented above are shown, but for the imaginary part of nonlinear susceptibility. In this case the anchor point is located at $\omega_1 = 1\,\mathrm{eV}$. From Fig. 8.7b, we observe that the precision of the data inversion is dramatically better when using SSKK rather than conventional K-K relations. The presence of the anchor point greatly reduces the errors of the estimation performed with conventional K-K relations in the energy range 0.9–1.4 eV.

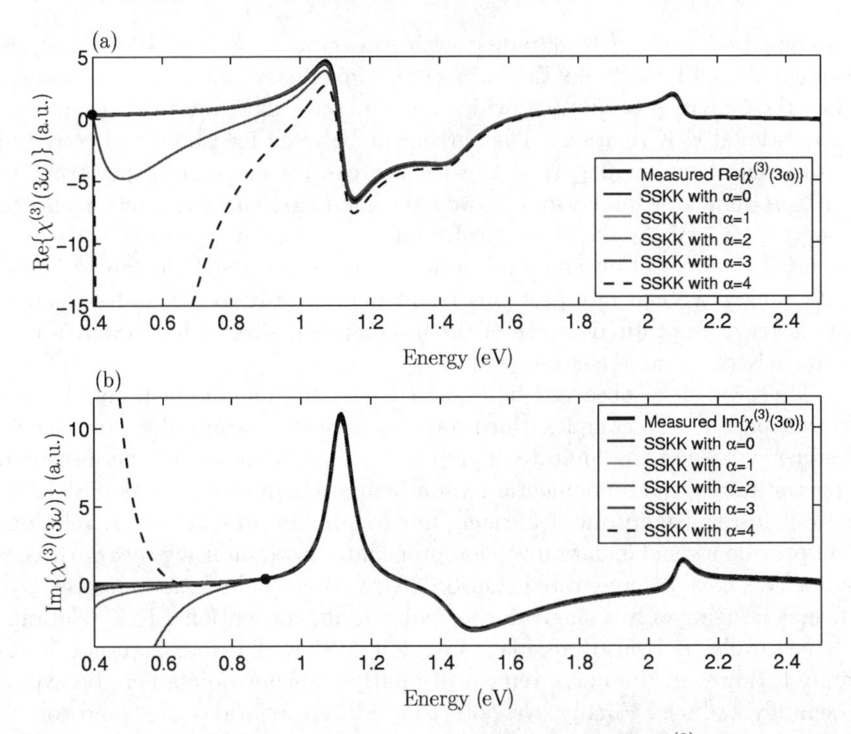

Fig. 8.8. Efficacy of SSKK relations in retrieving (**a**) $\mathrm{Re}\{\chi^{(3)}(3\omega)\}$ and (**b**) $\mathrm{Im}\{\chi^{(3)}(3\omega)\}$ over the full spectral range (•=anchor point). Reproduced from [33]

We then test the efficacy of the SSKK when the higher moments of susceptibility are considered, and we choose to consider the whole available spectral range $0.4 - 2.5\,\mathrm{eV}$, thus adopting a high data availability scenario [33].

The anchor point we select for the real part is at the lower boundary of the interval, thus simulating an experimental situation where we have information on quasi-static phenomena. In Fig. 8.8a, we present our results of the data inversion for the real part, which can be compared with those obtained with conventional K-K shown in Fig. 8.1a, which nevertheless were obtained after an optimization procedure. These should be interpreted more in terms of self-consistency analysis than in terms of pure data inversion. We can see that the SSKK outperform conventional K-K relations and provide virtually perfect data inversion for $0 \leq \alpha \leq 2$, while for $\alpha = 3$, the finite range causes disagreement between the measured and retrieved data for photon energy $\leq 0.9\,\mathrm{eV}$. In the $\alpha = 4$ case, for which the theory does not prescribe convergence, we have reasonable agreement only in the high-energy range.

In the case of the imaginary part, the anchor point is on the low-energy side of the main spectral feature, just out of the resonance. In Fig. 8.8b, we present the results of optical data inversion with SSKK relations for the imaginary part, which, with the same previously discussed caveats, should be compared with those obtained with conventional K-K relations and are presented in Fig. 8.1b. In the case of the imaginary part, we also observe that the SSKK procedure provides more precise data inversion than with conventional K-K relations. The agreement between measured and retrieved data is excellent for $0 \leq \alpha \leq 2$, with no relevant decrease of performance for increasing α, while for $\alpha = 3$, we have good agreement except for photon energy $\leq 0.7\,\mathrm{eV}$. In the theoretically nonconverging $\alpha = 4$ case, somewhat surprisingly we still obtain good performance of the SSKK for an extended data range. We can interpret this result heuristically as a manifestation of the faster asymptotic decrease of the integrands realized when SSKK rather than K-K relations are considered.

Thus, we have observed how an independent measurement of the unknown part of the complex third-order nonlinear susceptibility for a given frequency relaxes the limitations imposed by the finiteness of the measured spectral range, the fundamental reason being that in the obtained SSKK relations faster asymptotic decreasing integrands are present. SSKK relations can provide a reliable data inversion procedure based on using *measured data alone*. We have demonstrated that SSKK relations yield a more precise data inversion, using only a single anchor point, than conventional K-K relations.

Naturally, it is also possible to exploit MSKK if higher precision is required. However, the measurement of multiple anchor points may be experimentally tedious. Finally, we note that MSKK relations are valid for all holomorphic nonlinear susceptibilities of arbitrary order. As an example of such holomorphic third-order nonlinear susceptibilities, we mention those related to pump-and-probe nonlinear processes [129], previously analyzed in Sect. 6.4.

8.5 Estimates of the Truncation Error in Kramers-Kronig Relations

In order to approximately assess the truncation error in optical data inversion, we separate the contribution to the integration in the K-K relations related to the range $[\omega_{\min}, \omega_{\max}]$ covered by the experimental data, as follows:

$$\begin{aligned} &\mathrm{Re}\left\{\chi^{(n)}(n\omega)\right\} - \mathrm{Re}\left\{\chi^{(n)}(n\omega)_{\mathrm{tr}}\right\} = \\ &\qquad \frac{2}{\pi\omega^{2\alpha}}\mathrm{P}\int_0^{\omega_{\min}} \frac{\omega'^{2\alpha+1}\mathrm{Im}\left\{\chi^{(n)}(n\omega')\right\}}{\omega'^2-\omega^2}\mathrm{d}\omega' \\ &\qquad + \frac{2}{\pi\omega^{2\alpha}}\mathrm{P}\int_{\omega_{\max}}^{\infty} \frac{\omega'^{2\alpha+1}\mathrm{Im}\left\{\chi^{(n)}(n\omega')\right\}}{\omega'^2-\omega^2}\mathrm{d}\omega', \end{aligned} \tag{8.3}$$

$$\operatorname{Im}\left\{\chi^{(n)}(n\omega)\right\} - \operatorname{Im}\left\{\chi^{(n)}(n\omega)_{\mathrm{tr}}\right\} = \\ -\frac{2}{\pi\omega^{2\alpha-1}} \mathrm{P} \int_0^{\omega_{\min}} \frac{\omega'^{2\alpha}\operatorname{Re}\left\{\chi^{(n)}(n\omega')\right\}}{\omega'^2 - \omega^2} d\omega' \\ -\frac{2}{\pi\omega^{2\alpha-1}} \mathrm{P} \int_{\omega_{\max}}^{\infty} \frac{\omega'^{2\alpha}\operatorname{Re}\left\{\chi^{(n)}(n\omega')\right\}}{\omega'^2 - \omega^2} d\omega', \tag{8.4}$$

where the subscript tr refers to the truncated data inversion and $0 \leq \alpha \leq n$. We consider $\omega_{\min} \ll \omega \ll \omega_{\max}$ and approximate the real and imaginary parts of the susceptibility as follows:

$$\begin{aligned} \operatorname{Re}\left\{\chi^{(n)}(n\omega)\right\} &\sim \mathrm{A}\omega^{2a}, \quad \omega < \omega_{\min}, \\ \operatorname{Re}\left\{\chi^{(n)}(n\omega)\right\} &\sim \mathrm{B}\omega^{-2b}, \quad \omega > \omega_{\max}, \\ \operatorname{Im}\left\{\chi^{(n)}(n\omega)\right\} &\sim \mathrm{C}\omega^{2c+1}, \quad \omega < \omega_{\min}, \\ \operatorname{Im}\left\{\chi^{(n)}(n\omega)\right\} &\sim \mathrm{D}\omega^{-2d-1}, \quad \omega > \omega_{\max}, \end{aligned} \tag{8.5}$$

where $a, b, c, d \geq 0$ and $b, d \geq \alpha + 1$; and we have considered the symmetry properties of the real and imaginary part (7.3). By plugging the expansions (8.5) into (8.3) and (8.4) we obtain

$$\operatorname{Re}\left\{\chi^{(n)}(n\omega)\right\} - \operatorname{Re}\left\{\chi^{(n)}(n\omega)_{\mathrm{tr}}\right\} \sim \\ -\frac{2}{\pi}\left(\frac{\omega_{\min}}{\omega}\right)^{2\alpha+2} \operatorname{Im}\left\{\chi^{(n)}(n\omega_{\min})\right\} \frac{1}{2\alpha+2c+2} \\ +\frac{2}{\pi}\left(\frac{\omega_{\max}}{\omega}\right)^{2\alpha} \operatorname{Im}\left\{\chi^{(n)}(n\omega_{\max})\right\} \frac{1}{2d+1-2\alpha}, \tag{8.6}$$

$$\operatorname{Im}\left\{\chi^{(n)}(n\omega)\right\} - \operatorname{Im}\left\{\chi^{(n)}(n\omega)_{\mathrm{tr}}\right\} \sim \\ \frac{2}{\pi}\left(\frac{\omega_{\min}}{\omega}\right)^{2\alpha+1} \operatorname{Re}\left\{\chi^{(n)}(n\omega_{\min})\right\} \frac{1}{2\alpha+2a+1} \\ -\frac{2}{\pi}\left(\frac{\omega_{\max}}{\omega}\right)^{2\alpha-1} \operatorname{Re}\left\{\chi^{(n)}(n\omega_{\max})\right\} \frac{1}{2b+1-2\alpha}. \tag{8.7}$$

We observe that the corrections go to zero when $\omega_{\min} \to 0$ and $\omega_{\max} \to \infty$. Morever, these formulas explain why we can observe in Figs. 8.1–8.2 that the largest discrepancies between the retrieved and the actual data are located in the lower part of the energy spectrum and that the magnitude of the discrepancy consistently increases with decreasing values of ω. In particular, by substituting $a = 0$, $b = d = n{+}1$, $c = 1$, and $d = 2n{+}1$, as suggested by the results presented in Chap. 7 and by the anharmonic oscillator model [34,77], and selecting $\alpha = 0$,

$$\operatorname{Re}\left\{\chi^{(n)}(n\omega)\right\} - \operatorname{Re}\left\{\chi^{(n)}(n\omega)_{\mathrm{tr}}\right\} \sim$$
$$-\frac{1}{2\pi}\left(\frac{\omega_{\min}}{\omega}\right)^2 \operatorname{Im}\left\{\chi^{(n)}(n\omega_{\min})\right\} + \frac{2}{\pi(2n+3)}\operatorname{Im}\left\{\chi^{(n)}(n\omega_{\max})\right\}, \tag{8.8}$$

and

$$\operatorname{Im}\left\{\chi^{(n)}(n\omega)\right\} - \operatorname{Im}\left\{\chi^{(n)}(n\omega)_{\mathrm{tr}}\right\} \sim$$
$$\frac{2}{\pi}\frac{\omega_{\min}}{\omega}\operatorname{Re}\left\{\chi^{(n)}(n\omega_{\min})\right\} - \frac{2}{\pi(2n+3)}\frac{\omega}{\omega_{\max}}\operatorname{Re}\left\{\chi^{(n)}(n\omega_{\max})\right\}. \tag{8.9}$$

8.6 Sum Rules and Static Second-Order Nonlinear Susceptibility

Recently, it has been shown how to use the constraints imposed by the sum rules relevant to second-harmonic-generation susceptibility to derive efficient ways to estimate second-order static susceptibility $\chi^{(2)}_{ijk}(0;0,0)$ for semiconductors without a center of inversion [180]. Such a quantity has been difficult to estimate with ab initio as well as with semiempirical approaches because of the sensitivity to constituent parameters (energy differences and matrix elements) and because of the unavoidable truncation error due to the consideration of a finite set of eigenstates [175, 192, 193].

If we adopt a single-resonance simplified quantum model for solids, it is possible to write the second-harmonic-generation susceptibility as

$$\chi^{(2)}_{ijk} \sim \frac{\alpha^{(1)}_{ijk}}{\omega_0 - \omega - \mathrm{i}\gamma} + \frac{\alpha^{(2)}_{ijk}}{\omega_0 - 2\omega - \mathrm{i}\gamma} + \frac{\alpha^{(3)}_{ijk}}{(\omega_0 - \omega - \mathrm{i}\gamma)^2} + (\omega \to -\omega)^*, \tag{8.10}$$

where ω_0 is the resonance frequency; $\alpha^{(1)}_{ijk}$, $\alpha^{(2)}_{ijk}$, and $\alpha^{(3)}_{ijk}$ are tensorial parameters; and the last term indicates the terms obtained by changing ω into $-\omega$ and taking the conjugate of the first three terms. If we impose sum rules (7.18)–(7.20) on expression (8.10) with $k = 1$ and $n = 2$, we derive the values of parameters $\alpha^{(1)}_{ijk}$, $\alpha^{(2)}_{ijk}$, and $\alpha^{(3)}_{ijk}$, and find after lengthy calculations that the second-harmonic susceptibility can be expressed as a product of linear susceptibilities in the form of Miller's rule. The Miller empirical rule [194] generalizes the observation that in many materials, the second-harmonic-generation susceptibility in the transparency region can be expressed in terms of the product of suitable linear susceptibility functions:

$$\chi^{(2)}_{ijk}(2\omega;\omega,\omega) \sim \Delta_{ijk}\chi^{(1)}_{ii}(2\omega)\chi^{(1)}_{jj}(\omega)\chi^{(1)}_{kk}(\omega), \tag{8.11}$$

where Δ_{ijk} is a coupling constant, named Miller's Δ, relating nonlinear and the linear processes. Miller's Δ can be expressed as

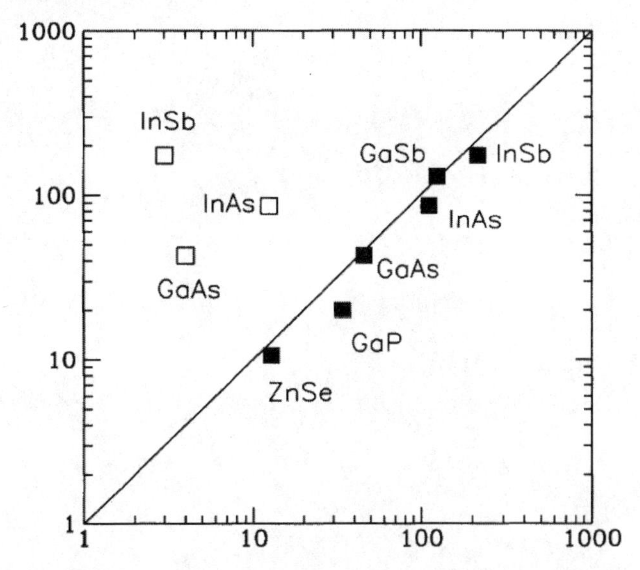

Fig. 8.9. Second-order static susceptibility for some semiconductors. *Abscissa*: theoretical calculations in units of 10^{-8} esu. *Ordinate*: experimental data in units of 10^{-8} esu. *Full squares*: data from [180]; *open squares*: data from [192]. Adapted from [180]

$$\Delta_{ijk} = \frac{1}{2e^3}\left(\frac{N}{V}\right)^{-2}\left\langle\frac{\partial^3 V(\boldsymbol{r})}{\partial r_i \partial r_j \partial r_k}\right\rangle_0 , \tag{8.12}$$

in agreement with the results obtained with the anharmonic oscillator model, when the quantum expectation value of the ground state substitutes for the evaluation at the equilibrium position [34, 77]. If we take the static limit of expression (8.11), we obtain

$$\chi^{(2)}_{ijk}(0;0,0) \sim \Delta_{ijk}\chi^{(1)}_{ii}(0)\chi^{(1)}_{jj}(0)\chi^{(1)}_{kk}(0). \tag{8.13}$$

Since the static linear susceptibility is a well-known and easily measurable quantity for every material, it is then possible to obtain an estimate of the second-order static susceptibility once Miller's Δ can be computed. As thoroughly explained in [180], this can be accomplished by considering the widely available pseudopotential functions and subtracting the electrons' Hartree screening. In Fig. 8.9, we show that such an approach, although based on a very simplified scheme, greatly outperforms much more sophisticated and time-expensive calculations, since it is based and focused on the fundamental properties of optical response functions.

9 Modified Kramers-Kronig Relations in Nonlinear Optics

9.1 Modified Kramers-Kronig Relations for a Meromorphic Nonlinear Quantity

In this chapter, we wish to develop a dispersion theory for nonlinear optics that overcomes some of the limitations of the tools introduced in the previous chapters. It is clearly stated by Toll [1] that causality is the primary reason for the existence of K-K relations in the field of linear optics. This is certainly true, but in the field of nonlinear optics, it cannot be taken for granted that causality is a necessary and sufficient condition for the validity of K-K relations, as thoroughly discussed in Chap. 6. Tokunaga et al. [195] have shown on grounds of the experimental and theoretical studies that K-K relations are of limited validity in pump-and-probe femtosecond time-resolved spectroscopy. According to them, only in some special cases, related to negative and positive time delays between pump-and-probe light pulses, are K-K relations valid, although obviously causality always works when the pump beam arrives before the probe in pump-and-probe experiments. Furthermore they also observed a switch of the sign of the K-K relations. They correctly argued that in general in the scheme of positive or zero delay, K-K relations are not valid at all.

As far as we know, the angular-frequency-dependent nonlinear susceptibility in any known case of nonlinear optical spectroscopy can be described either as a holomorphic or meromorphic quantity. In a sense, meromorphic nonlinear susceptibility presents a more general case than that of the holomorphic one. The reason is that in a mathematical sense, holomorphic nonlinear susceptibility can be considered a special case of meromorphic nonlinear susceptibility, which is a holomorphic function except at complex poles.

Below, we present results for a meromorphic nonlinear quantity, which may be nonlinear susceptibility, refractive index, or reflectivity, respectively, with the purpose of generalizing dispersion relations and sum rules. We assume the most general case of complex response function. Such a presumption implies that there is no assumption of specific symmetry of the real and imaginary parts of the nonlinear quantity [127]. Note that Tokunaga et al. [195] observed, simultaneously, real and imaginary parts of susceptibility, both of which were even functions at zero delay. In the context of the real response

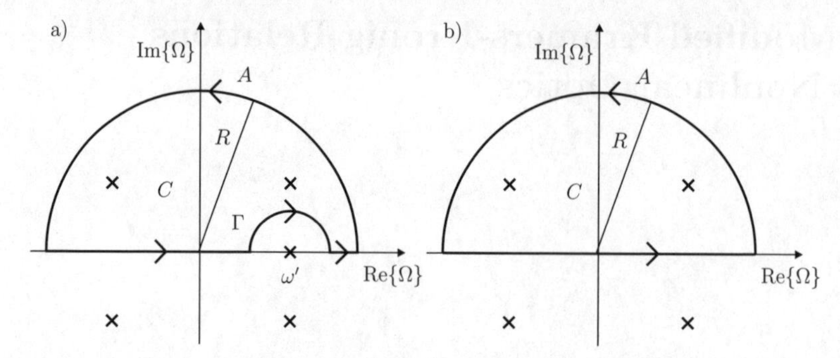

Fig. 9.1. Contour for the derivation of (**a**) the dispersion relations and (**b**) the sum rules (×=pole). Reproduced from [196]

function, the real and imaginary parts of nonlinear susceptibility always obey even and odd parity, respectively.

Now under the rather general assumption of considering a meromorphic function without parity, we define a function f, which is a nonlinear optical quantity as a function of a single angular frequency variable. Naturally, various angular frequencies and their combinations may be present, but, for the sake of simplicity in the following we indicate only one variable. This is natural since usually in multiple light beam experiments in nonlinear optics, only the wavelength of the probe beam is scanned, i.e. there is only one true variable, as discussed in the context of the Scandolo theorem in Chap. 6. The function f is assumed to be a complex function of real variable $f(\omega) = u(\omega) + \mathrm{i}v(\omega)$, where ω is the angular frequency. Next we use complex analysis and consider the function $f = f(\omega)$, as a meromorphic function of the complex frequency variable Ω. Next we consider a complex contour integration, as shown in Fig. 9.1a:

$$\oint_C \frac{f(\Omega)}{\Omega - \omega} \mathrm{d}\Omega = \mathrm{P} \int_{-R}^{R} \frac{f(\omega)}{\omega' - \omega} \mathrm{d}\omega' + \int_{\Gamma} \frac{f(\Omega)}{\Omega - \omega} \mathrm{d}\Omega + \int_{A} \frac{f(\Omega)}{\Omega - \omega} \mathrm{d}\Omega. \tag{9.1}$$

The next procedure is to let the radius R tend to infinity. The integration along the closed contour on the left hand-side of the equation gives, according to the theorem of residues, usually a nonzero contribution due to the poles located in the upper half plane. The first integral on the right-hand side of (9.1) is the integral that is the origin of Hilbert transforms. The second integral on the right hand-side in turn gives a nonzero contribution due to the residue theorem. Finally the last integral on the right-hand side of (9.1) is equal to zero (rigorous mathematical proof is presented in [188]). Finally, by separation of the real and imaginary parts, we get the results

$$u(\omega) = \frac{1}{\pi} \mathrm{P} \int_{-\infty}^{\infty} \frac{v(\omega')}{\omega' - \omega} \mathrm{d}\omega' - 2\mathrm{Re} \left\{ \sum_{\mathrm{Im}\{\Omega\}>0}^{\mathrm{poles}} \mathrm{Res} \left[\frac{f(\Omega)}{\Omega - \omega} \right] \right\} \tag{9.2}$$

and

$$v(\omega) = -\frac{1}{\pi}\mathrm{P}\int_{-\infty}^{\infty} \frac{u(\omega')}{\omega' - \omega}\mathrm{d}\omega' - 2\mathrm{Im}\left\{\sum_{\mathrm{Im}\{\Omega\}>0}^{\mathrm{poles}} \mathrm{Res}\left[\frac{f(\Omega)}{\Omega - \omega}\right]\right\}. \tag{9.3}$$

If there are no poles in the upper half plane, certainly then the residue terms in (9.2) and (9.3) are equal to zero. Thus we are dealing with a holomorphic quantity and Hilbert transforms hold. This means that the response function is real and, symmetry relations hold, which imply the validity of K-K relations. Fortunately, (9.2) and (9.3) can be written in slightly different forms by resolving the functions u and v into the sums of even and odd parts as follows:

$$u(\omega) = u_{\mathrm{even}}(\omega) + u_{\mathrm{odd}}(\omega), \tag{9.4}$$

$$v(\omega) = v_{\mathrm{even}}(\omega) + v_{\mathrm{odd}}(\omega). \tag{9.5}$$

Thus according to (9.2)–(9.5),

$$\begin{aligned} u(\omega) =& -\frac{2\omega}{\pi}\mathrm{P}\int_{0}^{\infty} \frac{v_{\mathrm{even}}(\omega')}{\omega'^2 - \omega^2}\mathrm{d}\omega' + \frac{2}{\pi}\mathrm{P}\int_{0}^{\infty} \frac{\omega v_{\mathrm{odd}}(\omega')}{\omega'^2 - \omega^2}\mathrm{d}\omega' \\ &- 2\mathrm{Re}\left\{\sum_{\mathrm{Im}\{\Omega\}>0}^{\mathrm{poles}} \mathrm{Res}\left[\frac{f(\Omega)}{\Omega - \omega}\right]\right\} \end{aligned} \tag{9.6}$$

and

$$\begin{aligned} v(\omega) =& \frac{2\omega}{\pi}\mathrm{P}\int_{0}^{\infty} \frac{u_{\mathrm{even}}(\omega')}{\omega'^2 - \omega^2}\mathrm{d}\omega' - \frac{2}{\pi}\mathrm{P}\int_{0}^{\infty} \frac{\omega' u_{\mathrm{odd}}(\omega')}{\omega'^2 - \omega^2}\mathrm{d}\omega' \\ &- 2\mathrm{Im}\left\{\sum_{\mathrm{Im}\{\Omega\}>0}^{\mathrm{poles}} \mathrm{Res}\left[\frac{f(\Omega)}{\Omega - \omega}\right]\right\}, \end{aligned} \tag{9.7}$$

where $\omega > 0$. From (9.6) and (9.7), we observe that on the right-hand side, there are combinations of K-K relations and in addition the residue term. The residue term can be problematic, because it requires knowledge of the complex function f of a complex variable. Because the poles are characterized by resonance points of the system, they can be estimated, provided that information on the transitional frequencies and the lifetimes of the electronic states is available. Then it is reasonable to try to construct the complex nonlinear function which is holomorphic almost everywhere except at the poles. In the most general case of meromorphic function, the existence of complex zeros with f is also allowed. Note that the dispersion theory above can also be applied to the powers of the function f, f^n, where n is an integer, and also to the appropriate moments, $\omega^k f^n$, where k is an integer.

Usually, the wavelength dependent spectrum in nonlinear optics is recorded. If we get information only on modulus $|f|$, then the real and imaginary parts can be retrieved both in linear and nonlinear optical spectroscopy by the maximum entropy method [15, 188], discussed in the next chapter. Then the calculation of the correlation function, in the case of a weak probe beam, may be based on the procedure presented by Remacle and Levine [127]. However, such a study is beyond the scope of this book.

9.2 Sum Rules for a Meromorphic Nonlinear Quantity

In the following, we consider some basic sum rules that are valid for meromorphic nonlinear quantities. If we set $\omega' = 0$ in (9.2) and (9.3), we get dc sum rules as follows:

$$u(0) = \frac{1}{\pi}\mathrm{P}\int_{-\infty}^{\infty} \frac{v(\omega')}{\omega'}\mathrm{d}\omega' - 2\mathrm{Re}\left\{\sum_{\mathrm{Im}\{\Omega\}>0}^{\mathrm{poles}} \mathrm{Res}\left[\frac{f(\Omega)}{\Omega}\right]\right\} \tag{9.8}$$

and

$$v(0) = -\frac{1}{\pi}\mathrm{P}\int_{-\infty}^{\infty} \frac{u(\omega')}{\omega'}\mathrm{d}\omega' - 2\mathrm{Im}\left\{\sum_{\mathrm{Im}\{\Omega\}>0}^{\mathrm{poles}} \mathrm{Res}\left[\frac{f(\Omega)}{\Omega}\right]\right\}. \tag{9.9}$$

If instead, we wish to write the static sum rules for the dispersion relations of (9.6) and (9.7) then the first integrals on the right-hand side of such equations vanish and only the odd parts of the functions contribute to the dc sum rules. dc sum rules constitute constraints that meromorphic nonlinear susceptibilities or memory functions have to obey.

Next we derive another set of sum rules, which may have practical utility at least in degenerate four-wave mixing spectroscopy. We exploit the complex contour integration shown in Fig. 9.1b and write the following equation:

$$\oint_C f(\Omega)d\Omega = \int_{-R}^{R} f(\omega')d\omega' + \int_A f(\Omega)d\Omega. \tag{9.10}$$

We let the radius R tend to infinity and note that the integral along arc A vanishes. With the aid of the theorem of residues, we get

$$I_1 = \mathrm{P}\int_{-\infty}^{\infty} u(\omega)\mathrm{d}\omega = -2\pi\mathrm{Im}\left\{\sum_{\mathrm{Im}\{\Omega\}>0}^{\mathrm{poles}} \mathrm{Res}\left[f(\Omega)\right]\right\} \tag{9.11}$$

and

$$I_2 = \mathrm{P}\int_{-\infty}^{\infty} v(\omega')\mathrm{d}\omega' = 2\pi\mathrm{Re}\left\{\sum_{\mathrm{Im}\{\Omega\}>0}^{\mathrm{poles}} \mathrm{Res}\left[f(\Omega)\right]\right\}. \tag{9.12}$$

In the case of the partition of even and odd functions, that is to say, (9.4) and (9.5) are exploited, then sum rules (9.11) and (9.12) will have changes on the left-hand sides of these equations. In other words, they involve integration only on the semi-infinite positive real axis, and the integrands involve only even functions. Sum rules such as those given by (9.11) and (9.12) constitute other constraints that the nonlinear meromorphic quantity has to fulfill. In the case of a holomorphic quantity, the only sum rule follows from (9.11), i.e.

$$\int_{-\infty}^{\infty} u(\omega')\mathrm{d}\omega' = 0. \tag{9.13}$$

Note that we have abandoned the principal value sign "P," since usually the integral exists as a conventional integral. Unfortunately, sum rules that involve residue terms, so far, have importance in testing theoretical models, whereas sum rules that deal directly with measured data have crucial practical importance in testing e.g. the validity of measured spectra. Therefore, we now take a step toward more practical sum rules, which, however, involve meromorphic nonlinear quantities. This is possible by integrating (9.2)–(9.3) as principal value integrals with respect to ω'. Then we find that

$$\begin{aligned} I_3 =& \mathrm{P}\int_{-\infty}^{\infty} u(\omega')\mathrm{d}\omega' = \mathrm{P}\int_{-\infty}^{\infty}\left[\frac{1}{\pi}\mathrm{P}\int_{-\infty}^{\infty}\frac{v(\omega)}{\omega-\omega'}\mathrm{d}\omega\right]\mathrm{d}\omega' \\ &- \mathrm{P}\int_{-\infty}^{\infty} 2\mathrm{Re}\left\{\sum_{\mathrm{Im}\{\Omega\}>0}^{\mathrm{poles}}\mathrm{Res}\left[\frac{f(\Omega)}{\Omega-\omega'}\right]\right\}\mathrm{d}\omega' \\ &= \mathrm{P}\int_{-\infty}^{\infty}\left[\frac{1}{\pi}\mathrm{P}\int_{-\infty}^{\infty}\frac{v(\omega)}{\omega-\omega'}\mathrm{d}\omega\right]\mathrm{d}\omega' - 2\pi\mathrm{Im}\left\{\sum_{\mathrm{Im}\{\Omega\}>0}^{\mathrm{poles}}\mathrm{Res}\left[f(\Omega)\right]\right\} \end{aligned} \tag{9.14}$$

and

$$\begin{aligned} I_4 =& \mathrm{P}\int_{-\infty}^{\infty} v(\omega')\mathrm{d}\omega' = \mathrm{P}\int_{-\infty}^{\infty}\left[-\frac{1}{\pi}\mathrm{P}\int_{-\infty}^{\infty}\frac{u(\omega)}{\omega-\omega'}\mathrm{d}\omega\right]\mathrm{d}\omega' \\ &- \mathrm{P}\int_{-\infty}^{\infty} 2\mathrm{Im}\left\{\sum_{\mathrm{Im}\{\Omega\}>0}^{\mathrm{poles}}\mathrm{Res}\left[\frac{f(\Omega)}{\Omega-\omega'}\right]\right\}\mathrm{d}\omega' \\ &= \mathrm{P}\int_{-\infty}^{\infty}\left[-\frac{1}{\pi}\mathrm{P}\int_{-\infty}^{\infty}\frac{u(\omega)}{\omega-\omega'}\mathrm{d}\omega\right]\mathrm{d}\omega' + 2\pi\mathrm{Re}\left\{\sum_{\mathrm{Im}\{\Omega\}>0}^{\mathrm{poles}}\mathrm{Res}\left[f(\Omega)\right]\right\}. \end{aligned} \tag{9.15}$$

Actually, principal value integration is needed because of the residue terms in (9.14)–(9.15) which diverge logarithmically. The last expressions in (9.14)–

(9.15) were obtained using the partial fraction of a meromorphic function [197] and assuming that the poles appearing in the upper half plane are of the first order. As far as we know, in all cases of nonlinear susceptibility, the order of the poles appearing in the upper half plane is $k = 1$, whereas the order of poles appearing in the lower half plane can be higher (see, e.g. [129]).

The calculation of the integral

$$I_5 = \mathrm{P}\int_{-\infty}^{\infty} 2\mathrm{Re}\left\{\sum_{\mathrm{Im}\{\Omega\}>0}^{\mathrm{poles}} \mathrm{Res}\left[\frac{f(\Omega)}{\Omega-\omega}\right]\right\}\mathrm{d}\omega \tag{9.16}$$

is based on the use of the partial fraction $c_{ik}/(\Omega - a_i)^k$, where a_{ik} are the poles of the function, k is a positive integer, and c_{ik} are complex constants. It is sufficient to demonstrate the calculation for one residue term because the other terms are obtained in a similar manner. Then we find that when $k = 1$,

$$\begin{aligned}
\mathrm{P}\int_{-\infty}^{\infty} \mathrm{Re}\left\{\frac{c_{i1}}{a_i-\omega}\right\}\mathrm{d}\omega &= \mathrm{P}\int_{-\infty}^{\infty} \mathrm{Re}\frac{\mathrm{Re}\{c_{i1}\}+\mathrm{iIm}\{c_{i1}\}}{\mathrm{Re}\{a_i\}-\omega+\mathrm{iIm}\{a_i\}}\mathrm{d}\omega \\
&= \mathrm{P}\int_{-\infty}^{\infty} \frac{\mathrm{Re}\{c_{i1}\}+[\mathrm{Re}\{a_i\}-\omega]+\mathrm{Im}\{c_{i1}\}\mathrm{Im}\{a_i\}}{(\mathrm{Re}\{a_i\}-\omega)^2+(\mathrm{Im}\{a_i\})^2}\mathrm{d}\omega \\
&= \mathrm{P}\int_{-\infty}^{\infty} \frac{\mathrm{Re}\{c_{i1}\}[\mathrm{Re}\{a_i\}-\omega]}{(\mathrm{Re}\{a_i\}-\omega)^2+(\mathrm{Im}\{a_i\})^2}\mathrm{d}\omega \\
&\quad +\mathrm{P}\int_{-\infty}^{\infty} \frac{\mathrm{Im}\{c_{i1}\}\mathrm{Im}\{a_i\}}{(\mathrm{Re}\{a_i\}-\omega)^2+(\mathrm{Im}\{a_i\})^2}\mathrm{d}\omega \\
&= \mathrm{P}\int_{-\infty}^{\infty} \frac{\mathrm{Im}\{c_{i1}\}\mathrm{Im}\{a_i\}}{(\mathrm{Re}\{a_i\}-\omega)^2+(\mathrm{Im}\{a_i\})^2}\mathrm{d}\omega = \pi\mathrm{Im}\{c_{i1}\}.
\end{aligned} \tag{9.17}$$

Now if we compare (9.11), (9.12), (9.14), and (9.15), we find that [196]

$$\mathrm{P}\int_{-\infty}^{\infty}\left[\mathrm{P}\int_{-\infty}^{\infty}\frac{v(\omega')}{\omega'-\omega}\mathrm{d}\omega'\right]\mathrm{d}\omega = 0 \tag{9.18}$$

and

$$\mathrm{P}\int_{-\infty}^{\infty}\left[\mathrm{P}\int_{-\infty}^{\infty}\frac{u(\omega')}{\omega'-\omega}\mathrm{d}\omega'\right]\mathrm{d}\omega = 0. \tag{9.19}$$

The order of the double integration can be changed only upon the assumption of uniform convergent integrals. In such a case it is possible to integrate the function $(\omega'-\omega)^{-1}$ separately in (9.14)–(9.15). King [198] used such a strategy to derive sum rules for optical constants in linear optics.

10 The Maximum Entropy Method: Theory and Applications

10.1 The Theory of the Maximum Entropy Method

In the measurement of an optical power spectrum, the intensity distribution of light $I(\omega)$ is proportional to the squared modulus $|f(\omega)|^2$ of a complex function $f(\omega)$ evaluated for $\omega \in \mathbb{R}$. For instance, in reflection spectroscopy, the intensity reflectance $R(\omega) = |r(\omega)|^2$ is measured. Typically, while only the modulus of $f(\omega)$ can actually be measured, it is necessary to know the complex function $f(\omega) = |f(\omega)| \exp[\mathrm{i}\theta(\omega)]$ itself, including also the phase $\theta(\omega)$ in order to have a complete picture of the properties of the sample under investigation. A new phase retrieval approach was proposed by Vartiainen et al. [199]. The method uses the maximum entropy model or method (MEM). Its basis can be found in the Burg study about the calculation of a power spectrum of a finite time series [182]. The Burg idea was to choose the spectrum that corresponds to the most random time series, whose autocorrelation function agrees with a set of known values. This approach leads to a model for the power spectrum by maximizing the entropy of the corresponding time series, which is the reason for the name MEM. With its close relation to the concept of maximum entropy, this theory has also been used in optical spectroscopy, e.g. as a line-narrowing procedure [200, 201].

The applicability of MEM as a phase retrieval procedure has been verified for linear reflectance from solid [183, 202, 203] and liquid [204] phases. In the study of [204], the reflectometric data obtained on liquids from the process industry and subsequent comparison with other spectral devices and data analysis methods indicated the correct functioning of MEM analysis.

In an optically nonlinear medium, MEM can be applied to phase retrieval from the measured modulus of nonlinear susceptibility, in the anti-Stokes Raman scattering spectrum [205], the third-harmonic wave from polysilane [206], sum frequency generation spectroscopy [207, 208], meromorphic total susceptibility [209], degenerate nonlinear susceptibility from Maxwell Garnett nanocomposites [210], and reflectivity of nonlinear Bruggeman liquids [71].

The merit of MEM is that it does not require, in principle, determination of the intensity over the whole electromagnetic spectrum, but only of the region $\omega_1 \leq \omega \leq \omega_2$ of interest. As well as intensity data, additional information about a given medium is required in order to determine its complex optical properties. Such information at anchor points commonly comprises the real and/or imaginary parts of the complex quantity of the medium determined at a frequency within the considered range $\omega_1 \le \omega \le \omega_2$.

In practice, the phase retrieval procedure using the MEM requires fitting a power spectrum, e.g. an experimental reflectance R, by its maximum entropy model (the derivation can be found in [211]) given by

$$R(\nu) = \frac{|\beta|^2}{|A_M(\nu)|^2}, \tag{10.1}$$

where $A_M(\nu) = 1 + \sum_{m=1}^{M} a_m \exp(-\mathrm{i}2\pi m\nu)$ is a MEM polynomial given by the MEM coefficients a_m and by the normalized frequency ν. The latter is defined by the measurement range $[\omega_1, \omega_2]$ of $R(\omega)$ as $\nu = (\omega - \omega_1)/(\omega_2 - \omega_1)$. All the unknown MEM coefficients a_m and $|\beta|^2$ are obtained from a linear Toeplitz system:

$$\begin{pmatrix} C(0) & C(-1) & \cdots & C(-M) \\ C(1) & C(0) & \cdots & C(1-M) \\ \vdots & \vdots & \ddots & \vdots \\ C(M) & C(M-1) & \cdots & C(0) \end{pmatrix} \begin{pmatrix} 1 \\ a_1 \\ \vdots \\ a_M \end{pmatrix} = \begin{pmatrix} |\beta|^2 \\ 0 \\ \vdots \\ 0 \end{pmatrix}, \tag{10.2}$$

where the autocorrelation function $C(t)$ is defined by a Fourier transform of $R(\nu)$ as

$$C(t) = \int_0^1 R(\nu) \exp[\mathrm{i}2\pi t\nu] \mathrm{d}\nu. \tag{10.3}$$

The phase retrieval is then based on deriving the MEM for the complex reflectivity $r(\nu) = R^{1/2}(\nu) \exp[\mathrm{i}\theta(\nu)]$. This is realized by defining a MEM phase, $\psi(\nu)$, connected to the MEM polynomial as $A_M(\nu) = |A_M(\nu)| \exp[\mathrm{i}\psi(\nu)]$ and using (10.1) to get

$$r(\nu) = \frac{|\beta| \exp[\mathrm{i}\theta(\nu)]}{|A_M(\nu)|} = \frac{|\beta| \exp[\mathrm{i}(\theta(\nu) - \psi(\nu))]}{|A_M(\nu)| \exp[-\mathrm{i}\psi(\nu)]} = \frac{|\beta| \exp[\mathrm{i}\phi(\nu)]}{A_M^*(\nu)}. \tag{10.4}$$

Equation (10.4) introduces an error phase, $\phi(\nu) = \theta(\nu) - \psi(\nu)$, which provides a slowly varying background to $\theta(\nu)$, whereas $\psi(\nu)$ has the same spectral features as $\theta(\nu)$. Since the MEM phase ψ is obtained by the MEM fitting of R, the problem of finding the phase $\theta(\nu)$ is now reduced to a problem of finding the error phase. The idea in this is that typically $\phi\theta(\nu)$ is a much simpler function than $\theta(\nu)$ [183]. Thus, any additional information on $r(\nu)$ that can be used to obtain discrete values of $\theta(\nu_l)$ at frequencies ν_l, $l = 0, 1, \ldots, L$, can be used to obtain a good estimate for $\phi(\nu)$ by, e.g., a polynomial interpolation as [203]

$$\widehat{\phi}(\nu) = B_0 + B_1\nu + \cdots + B_L\nu^L = \sum_{l=0}^{L} B_l \nu^l. \tag{10.5}$$

The error phase is usually a slowly varying function and in favorable cases, only one or two anchor points are needed, i.e. the optimum degree of the polynomial is low. Brun et al. [212] presented an optimization method in order to smoothen the error phase of a reflection spectrum. Vartiainen et al. [184] derived sum rules in testing nonlinear susceptibility obtained using the maximum entropy model.

10.2 The Maximum Entropy Method in Linear Optical Spectroscopy

10.2.1 Phase Retrieval from Linear Reflectance

In linear reflection spectroscopy, the optical constants of a medium can be obtained usually by ellipsometry. However, if the optical constants are required across a relatively broad wavelength region, the conventional reflectometric measurement performed by scanning the wavelength of the incident light is usually a more appropriate method of measurement. In the case of normal incidence, the polarization of the incident light is not relevant, whereas in the case of oblique incidence, the nature of light polarization has to be properly taken care. In the case of an ideal surface, reflection is governed by Fresnel's equations. In the case of normal incidence, we measure the reflectance, which according to the (4.30) and (4.32) can be expressed as

$$R(\omega) = \left| \frac{N(\omega) - 1}{N(\omega) + 1} \right|^2 . \tag{10.6}$$

Thus if we can resolve the complex reflectivity, then the real refractive index and the extinction coefficient are obtained using relations

$$\eta(\omega) = \mathrm{Re}\left\{ \frac{1 + r(\omega)}{1 - r(\omega)} \right\} \quad , \quad \kappa(\omega) = \mathrm{Im}\left\{ \frac{1 + r(\omega)}{1 - r(\omega)} \right\} . \tag{10.7}$$

In Fig. 10.1a, we show the reflectance curve of a KCl crystal. It has been possible to derive the complex reflectivity from (10.4) by inserting the reflectance data into (10.1) and considering only one anchor point, selected on the basis that there is usually a wavelength for insulators, far from resonances, such that $\mathrm{Im}\{r(\omega)\}$ vanishes. In Fig. 10.1b,c we show the real refractive index and the extinction coefficient calculated from the real and imaginary parts of reflectivity by using the MEM procedure and by applying (10.7).

The wavelength-dependent reflectance of liquids is usually measured by a prism reflectometer, such as the device shown in Fig. 10.2, developed by Räty et al. [214]. In that device, the angle of incidence φ_i is oblique but fixed. This means that it is necessary to take care of the polarization degree of the incident light. Since the interface between the face of a prism and a liquid is ideal, Fresnel's equations for oblique incidence can be exploited in the analysis of the spectra. The usual way is to measure the reflectance for TE- or TM-polarized light and use the MEM. In such a case, the intensity reflectances

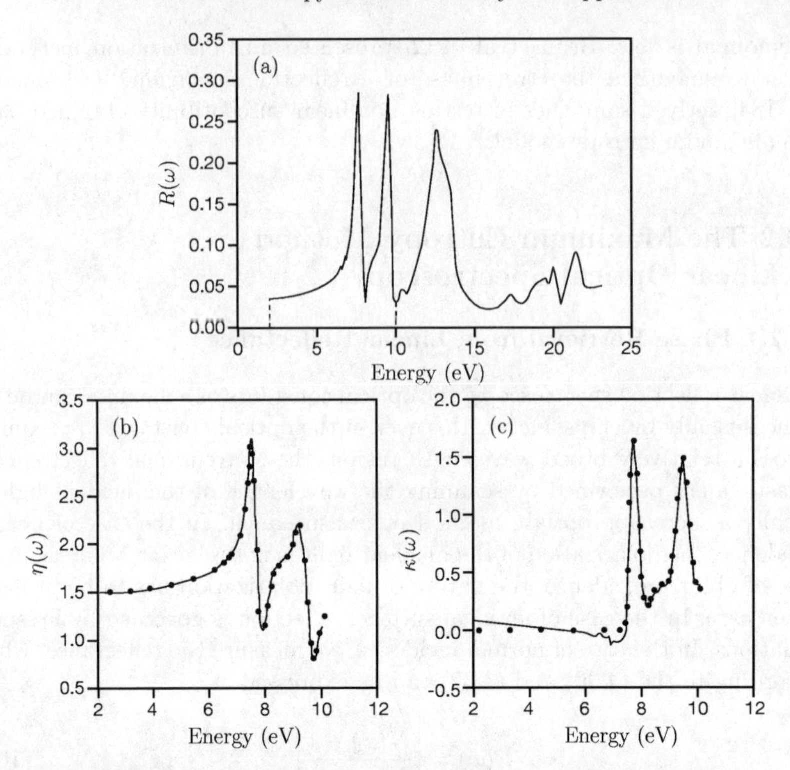

Fig. 10.1. (**a**) Reflectance of a KCl crystal as a function of energy. Exact (*dots*) and calculated (*solid lines*); (**b**) real refractive index η and (**c**) extinction coefficient κ of KCl. The calculations were done using the reflectancespectrum within the energy range from 2 to 10 eV. Reproduced from [199]

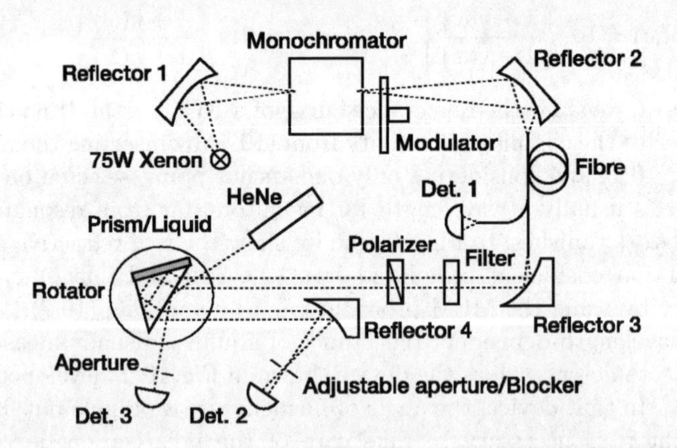

Fig. 10.2. A schematic optical layout of the reflectometer. Reproduced from [213]

are those given by (4.38) and (4.39). As an example, we consider the case of lignin diluted in water. Lignin is one of the main components of wood fiber and has particular importance in some industrial sectors such as the paper industry. Lignin absorbs UV light, which causes the aging of white paper, since it turns yellowish after long exposure to sunlight. In order to prevent the aging process, lignin is removed from the pulp in the papermaking process. In Fig. 10.3a, we show a reflectance curve for TE-polarized light, measured using the reflectometer of Fig. 10.2. Now the task is to find the phase of r_{TE}. For this purpose, we set

$$\sqrt{N^2 - \sin^2 \varphi_{\mathrm{i}}} = a + ib. \tag{10.8}$$

This means that we can write

$$r_{\mathrm{TE}} = |r_{\mathrm{TE}}| \exp^{[\mathrm{i}\theta]} = \sqrt{R_{\mathrm{TE}}} \exp^{[\mathrm{i}\theta]} = c + \mathrm{i}d = \frac{\cos\varphi_{\mathrm{i}} - (a+ib)}{\cos\varphi_{\mathrm{i}} + (a+ib)}. \tag{10.9}$$

Now we can separate the real and imaginary parts in (10.9) as follows:

$$a = \frac{1-e}{1+e} \cos\varphi_{\mathrm{i}}, \tag{10.10}$$

$$b = \frac{-2d}{(1+d)(1+e)} \cos\varphi_{\mathrm{i}}, \tag{10.11}$$

where

$$e = c + \frac{d^2}{1+c}. \tag{10.12}$$

With the aid of (10.8)–(10.11), the complex refractive index can be calculated once the phase θ has been resolved from (4.38) by the MEM. In the case of lignin solution, the anchor points were obtained by measuring the reflectance of TE-polarized light adopting fixed wavelengths but scanning the angle of incidence. Then the information on the chosen anchor points is obtained by an optimization procedure, which is based on minimizing the sum

$$S = \sum_{\theta} \left[R_{\mathrm{m}}(\varphi_{\mathrm{i}}) - R_{\mathrm{t}}(\varphi_{\mathrm{i}}, \eta_{\mathrm{prism}}, N)\right]^2 , \tag{10.13}$$

where R_{m} is the measured reflectance; R_{t} is the theoretical reflectance, obtained from (4.38); and η_{prism} is the real refractive index of the prism.

Usually the optimum degree of the polynomial expression in (10.5) is relatively low. The ideal case occurs when the error phase is given by linear estimation. For the purpose of finding good linear estimations, we have found that *squeezing* the measured spectrum is a useful procedure. The squeezed spectrum is obtained as follows:

$$R^{\mathrm{sq}}(\nu; K) \equiv R(\omega_1) \quad , \quad 0 \le \nu < z_K(\omega_1), \tag{10.14}$$

$$R^{\mathrm{sq}}(\nu; K) \equiv R(\omega) \quad , \quad z_K(\omega_1) \leq \nu \leq z_K(\omega_2), \tag{10.15}$$

$$R^{\mathrm{sq}}(\nu; K) \equiv R(\omega_2) \quad , \quad z_K(\omega_2) < \nu \leq 1, \tag{10.16}$$

where

$$z_K(\omega) = (2K+1)^{-1}\left(\frac{\omega - \omega_1}{\omega_2 - \omega_1} + K\right) \tag{10.17}$$

and

$$\nu = \frac{z_K(\omega) - z_K(\omega_1)}{z_K(\omega_2) - z_K(\omega_1)}. \tag{10.18}$$

By choosing $K = 0$, we restore the original measured spectrum, whereas when $K = 2$, the spectrum is squeezed from the interval $\nu \in [0, 1]$ into the interval $\nu \in [0.4, 0.6]$. In Fig. 10.3b,c, we show the real refractive index and the extinction coefficient obtained for the lignin solution by applying the MEM procedure described above and considering two anchor points at wavelengths 280 and 400 nm. The results for the complex refractive index are in good agreement with those obtained by other optical techniques [204]. The MEM procedure has also been successful in the estimation of the effective complex refractive index of plastic pigments used in paint and paper [215].

Finally, we mention that simulations on magnetoreflectance [216], using the squeezing procedure along with the MEM, have shown that it is possible to extract the complex refractive index of left- and right-hand circularly polarized light.

10.2.2 Study of Surface Plasmon Resonance

The excitation of a plasmon, which is a quantized collective oscillation of electrons, is possible by introducing a fluctuation in the charge density of a metal. Surface plasmons exist at the boundary of the metal and can be produced in some cases by using an external electric field. Let us first consider the *volume* plasmon using the classical Drude–Lorentz oscillator model [12, 14, 15, 17, 35, 38, 77]. As opposed to the case of insulators, when metals are considered, there is no restoring force keeping the conduction electrons confined in the vicinity of the nucleus. Then the complex dielectric function $\varepsilon_{\mathrm{mr}}$ of a metal is

$$\mathrm{Re}\{\varepsilon_{\mathrm{mr}}\} = 1 - \omega_{\mathrm{p}}^2 \frac{1}{\omega^2 + \gamma^2}, \tag{10.19}$$

$$\mathrm{Im}\{\varepsilon_{\mathrm{mr}}\} = \omega_{\mathrm{p}}^2 \frac{\gamma}{\omega(\omega^2 + \gamma^2)}, \tag{10.20}$$

which can de derived from the general Drude-Lorentz susceptibility by setting the resonance angular frequency $\omega_0 = 0$ and considering the plasma frequency ω_{p} introduced in (3.43). As can be easily seen by inspection of

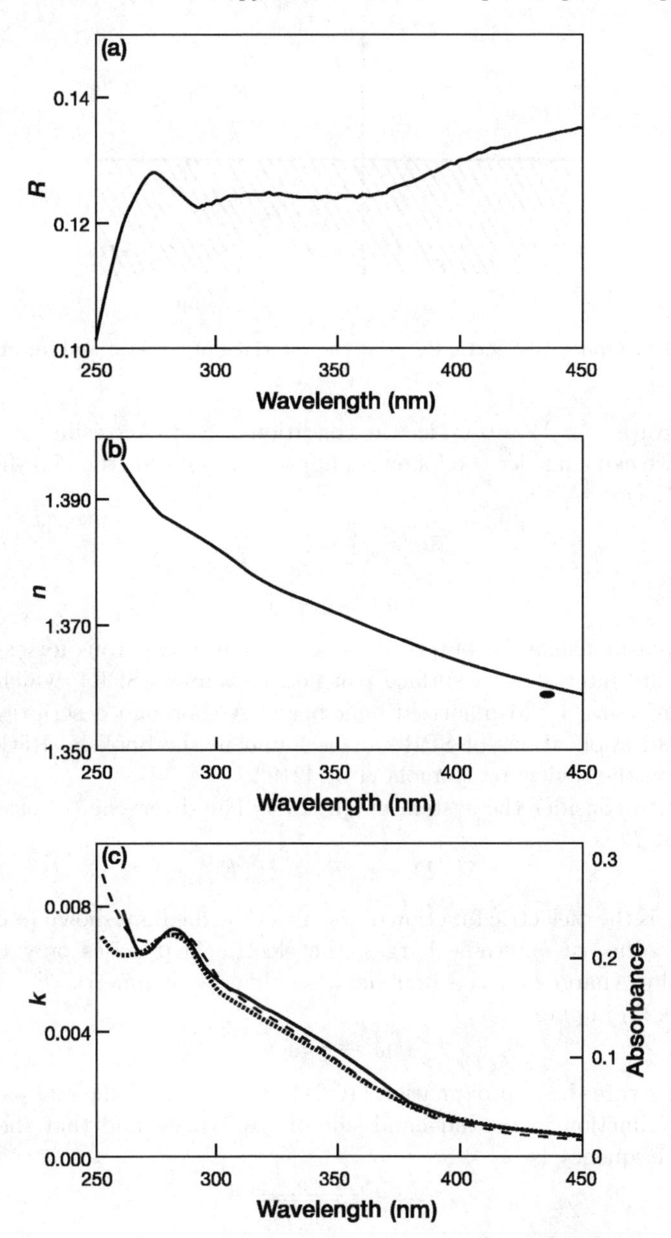

Fig. 10.3. (**a**) Experimental reflectance curve for a water–lignin solution, (**b**) real refractive index, and (**c**) extinction coefficient calculated using the MEM method. The reference measurements are denoted as follows: Abbe refractometer (*dot*), transmission (*dotted line*), and Attenuated Total Reflection (ATR) method (*dashed line*) (scale on the right side of the graph). Reproduced from [217]

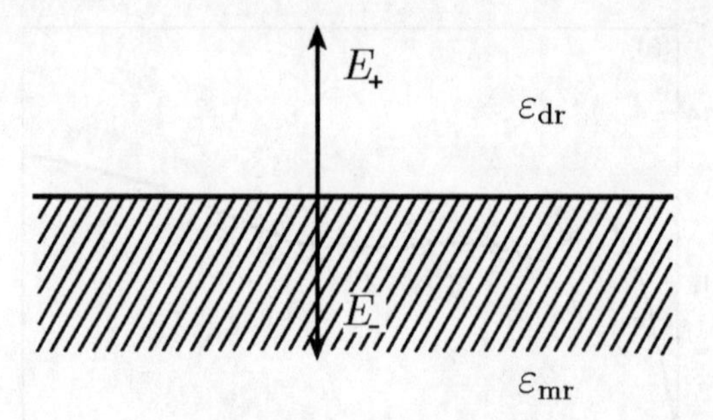

Fig. 10.4. Opposite electric fields at the interface of a metal and an insulator

(10.19), $\mathrm{Re}\{\varepsilon_{mr}(\omega_p)\} \sim 0$ under the condition $\omega \gg \gamma$. According to (10.19)–(10.20), we can consider the following approximations for the high-frequency range:

$$\mathrm{Re}\{\varepsilon_{mr}\} \sim 1 - \frac{\omega_p^2}{\omega^2}, \tag{10.21}$$

$$\mathrm{Im}\{\varepsilon_{mr}\} \ll 1. \tag{10.22}$$

Volume plasmons can be observed using a beam of electrons for excitation. Here we are interested in surface plasmon resonance (SPR), which can be excited by using a TM-polarized light beam. A thorough description of the theory and applications of SPR can be found in the book by Räther [218] and also in the review by Homola et al. [219].

We then consider the system of Fig. 10.4. The divergence of electric displacement $\boldsymbol{D}$ is

$$\nabla \cdot \boldsymbol{D} = \varepsilon_{dr} E_+ - \varepsilon_{mr} E_-, \tag{10.23}$$

where ε_{dr} is the dielectric function of the dielectric medium shown in Fig. 10.4. In the absence of external charges, the electric field arises only from the polarization charges on the boundary, so that by symmetry, $E_+ = -E_-$. Then (10.23) implies that

$$\varepsilon_{mr} = -\varepsilon_{dr}. \tag{10.24}$$

If we substitute the approximation (10.21) and (10.21) of the real part of the dielectric function in the left-hand side of (10.24), we find that the surface plasmon frequency is

$$\omega_S = \frac{\omega_p}{\sqrt{\varepsilon_{dr} + 1}}. \tag{10.25}$$

It follows from Maxwell's equations that when a light beam has a TM-polarized light component, only that component can generate a surface plasma wave [220]. The oscillation of surface charge fluctuations cannot ordinarily be excited by light. Kretschmann and Räther proposed the exploitation of a prism and a thin metallic film for surface plasma wave generation. Then the observation of reflectance in the ATR mode, using e.g. the reflectometer of Fig. 10.2, provides information about the resonance. That is to say, at a specific angle of incidence, the reflectance has a dip due to the fact that the light beam is coupled more effectively to the metal film. The resonance angle is larger than the critical angle, and it depends on the complex dielectric function of the metal film, on the optical properties of the liquid (or gas) to be studied, and on the refractive index of the prism. The reflectance can be derived by inspection of the multiple light reflection in an ambient–film–substrate system [221]. The expression of the reflectance is as follows:

$$R(\varphi_i) = \left| \frac{r_{\mathrm{pm}}(\varphi_i) + r_{\mathrm{ml}}(\varphi_i) \exp\left[2ik_z(\varphi_i)d\right]}{1 + r_{\mathrm{pm}}(\varphi_i) r_{\mathrm{ml}}(\varphi_i) \exp\left[2ik_z(\varphi_i)d\right]} \right|^2 , \tag{10.26}$$

where r_{pm} is the electric field reflectance at the prism–metal film interface, r_{ml} is the corresponding reflectance at the metal–liquid interface, d is the thickness (typically around 50 nm) of the metal film, and k_z is the scalar component of the wave vector normal to the metal film surface. The electric field reflectances are

$$r_{\mathrm{pm}} = \frac{k_{z,\mathrm{prism}}/\varepsilon_{\mathrm{prism,r}} - k_{z,\mathrm{m}}/\varepsilon_{\mathrm{mr}}}{k_{z,\mathrm{prism}}/\varepsilon_{\mathrm{prism,r}} + k_{z,\mathrm{m}}/\varepsilon_{\mathrm{mr}}} , \tag{10.27}$$

$$r_{\mathrm{ml}} = \frac{k_{z,\mathrm{m}}/\varepsilon_{\mathrm{m,r}} - k_{z,\mathrm{liq}}/\varepsilon_{\mathrm{liq,r}}}{k_{z,\mathrm{m}}/\varepsilon_{\mathrm{m,r}} + k_{z,\mathrm{liq}}/\varepsilon_{\mathrm{liq,r}}} , \tag{10.28}$$

where $\varepsilon_{\mathrm{prism,r}}$, $\varepsilon_{\mathrm{m,r}}$, and $\varepsilon_{\mathrm{liq,r}}$ are the dielectric functions of the prism, the metal film, and the liquid, respectively. The wave number is defined as follows:

$$k_{zj} = \left[\varepsilon_{jr} \left(\frac{\omega}{c} \right)^2 - k_x^2 \right]^{1/2} , \tag{10.29}$$

where

$$k_x = \eta_{\mathrm{prism}} \frac{\omega}{c} \sin \varphi_i . \tag{10.30}$$

In the vicinity of the dip, the reflectance can be approximated by using a Lorentzian line model [222, 223], which makes it possible to estimate the dielectric function of the metal film, its thickness, and the uncertainties in these estimates [224].

SPR for material research has turned out to be a very sensitive technique to detect small changes in the refractive index of gaseous [225] and liquid phases. Nowadays, there are various measurement techniques, which employ, for instance, a grating configuration instead of a prism, in the detection of physicochemical changes in media based on SPR. Thus SPR has proved to be a valuable tool, for instance, in the analysis of dynamic biological interactions [226,227]. A popular device, which is based on Kretschmann's configuration, exploits a flow cell that introduces on analyte solution, which passes through the thin metal film of the prism [228]. The metal film is usually polymer-coated and the adsorption of proteins onto the polymer film is monitored by detection of time-dependent SPR.

Matsubara et al. [229] introduced a device, which makes use of a convergent light beam such that there is no need for rotating the prism. This technique has also been exploited in commercial devices. The limitation of such a method is that only a relatively narrow refractive index range can usually be covered.

There is another technique also based on Kretschmann's configuration but characterized by keeping the angle of incidence fixed and scanning instead the wavelength of the light [230], which makes it possible to attain wavelength-dependent complex refractive index of liquids in the SPR measurement mode. Saarinen et al. [231] developed an analytical method, which allows calculations of the complex refractive index of a liquid using SPR reflectance by scanning the wavelength of light at a fixed angle of incidence. In that scheme, the unknown wavelength-dependent complex dielectric function of the liquid is obtained from the formulas above as follows:

$$\varepsilon_{\mathrm{liq,r}} = \varepsilon_{\mathrm{prism,r}} \frac{k_{z,\mathrm{liq}}(1-r_{\mathrm{pm}})(1-r_{\mathrm{ml}})}{k_{z,\mathrm{prism}}(1+r_{\mathrm{pm}})(1+r_{\mathrm{ml}})} . \tag{10.31}$$

After some algebra (see the detailed derivation presented in the Appendix of [231]), (10.31) can be put to into a more practical form:

$$\varepsilon_{\mathrm{liq,r}} = \frac{1}{2}\left[C + (C^2 - 4C\varepsilon_{\mathrm{prism,r}} \sin^2 \varphi_i)\right] , \tag{10.32}$$

where

$$C = \frac{2\pi}{\lambda}\left[\frac{\varepsilon_{\mathrm{prism,r}}(1-r_{\mathrm{pm}})(1-r_{\mathrm{ml}})}{k_{z,\mathrm{prism}}(1+r_{\mathrm{pm}})(1+r_{\mathrm{ml}})}\right] . \tag{10.33}$$

The method is based on the application of the MEM phase retrieval procedure. In other words, the phase of the complex reflectance appearing inside the modulus, on the right-hand side of (10.26), is calculated from the wavelength-dependent R. As an example of such a calculation, we present a simulation of the optical properties of MG nanoparticles in a water matrix [232]. The results of such calculations are shown in Fig. 10.5. The deep dips in Fig. 10.5a present reflectance minima due to SPR for three different volume fractions of nanoparticles. There are also shallow dips present in the vicinity of 2.3 eV, which are due to absorption. In the simulation, the dielectric function of water was known as a function of energy, and the dielectric function of the inclusions was assumed to obey a single Lorentzian resonance, as follows:

$$\varepsilon_{\mathrm{i}}(\omega) = \varepsilon_\infty + \frac{\omega_p^2}{\omega_0^2 - \omega^2 - \mathrm{i}\gamma\omega} , \tag{10.34}$$

where ε_∞ is the high-frequency dielectric function and the other symbols have their usual meanings. The MG model comes into the picture through the application of (3.59). The ME phases, calculated by assuming five anchor points, are shown in Fig. 10.5b. The corresponding real and imaginary

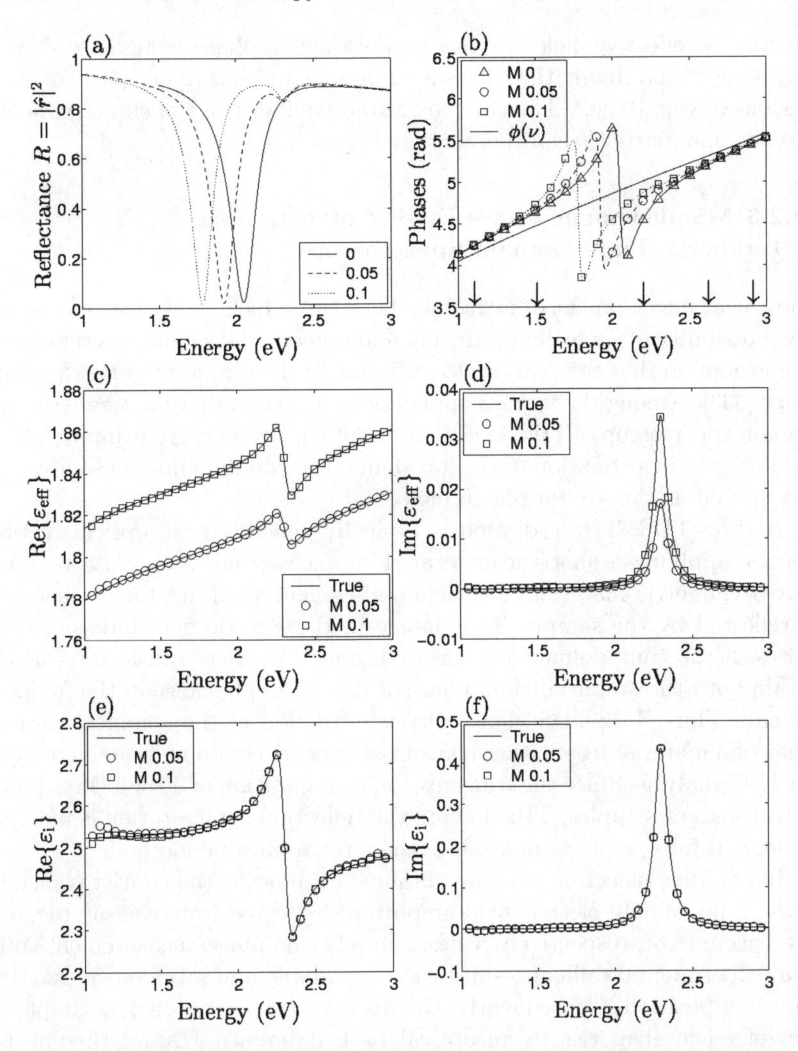

Fig. 10.5. (**a**) SPR reflectances of an MG liquid with three different volume fractions of inclusions, (**b**) the reconstructed true phases of reflectances with their MEM estimates (symbols), (**c**) the real, and (**d**) imaginary parts of the effective dielectric function of the medium (*solid lines*) along with their MEM estimates (*symbols*), (**e**) the real, and (**f**) imaginary parts of the inclusions (*solid lines*) along with their MEM estimates (*symbols*). The *arrows* in (**b**) show the frequencies at which the refractive index of the host liquid is assumed to be known in order to estimate the ME phases, which are represented by solid overlapping, almost linear, lines in (**b**). The angle of incidence is 70°, and the parameters are $\varepsilon_\infty = 2.5$, $\omega_{\mathrm{p}}^2 = 0.1\,\mathrm{eV}^2$, $\omega_0 = 2.3\,\mathrm{eV}$, $\Gamma = 0.1\,\mathrm{eV}$. Reproduced from [232]

parts of the effective dielectric function of the two-phase system are shown in Fig. 10.5c,d, and finally the corresponding real and imaginary parts of the inclusions in Fig. 10.5e,f. The complex refractive index of the effective medium and the nanoparticles can be calculated as well.

10.2.3 Misplacement Phase Error Correction in Terahertz Time-Domain Spectroscopy

The recent development of ultrashort laser pulse technology has opened up a new possibility of optically producing and detecting electromagnetic waves in a time domain that corresponds to radiation in the terahertz (THz) frequency range [233]. Among the various applications of THz radiation, terahertz time-domain spectroscopy (THz-TDS) [234–236] has aroused great interest for its capability in investigation of the intraband electron dynamics of semiconductors as well as the excitations in molecular systems.

In THz-TDS, THz radiation is typically generated via optical rectification by applying a short near-infrared subpicosecond laser pulse to a non-centrosymmetric material. The light pulses generated are then transmitted or reflected by the sample. The change in the electric field intensity of the pulses in the time domain is measured, and a Fourier transform is used to obtain both the amplitude and phase of the THz light pulse in the frequency domain. Thus, THz-TDS allows direct extraction of the complex refractive index of a material from either transmission or reflection measurements without K-K analysis. Since the transmission configuration of THz-TDS is limited to transparent samples, THz-TDS in the reflection configuration is more versatile, and for opaque samples, it is the only applicable method.

In practice, reflection data are obtained by dividing the Fourier transforms of the time-domain electric field amplitudes reflected from the sample and a reference mirror, respectively. Unfortunately, the phase measurement suffers from experimental difficulties in placing a sample and a reference exactly in the same position. Consequently, the displacement between the sample and the reference gives rise to an optical path difference δL and thereby to a relatively large frequency-dependent error into the obtained phase. Accordingly, the correction of phase error is a crucial point in THz spectroscopy, and various experimental techniques have been reported for minimizing the error [234, 236–238]. In contrast to these experimental techniques, a numerical method utilizing the MEM has been proposed to reveal the phase error in THz-TDS and to simplify the experimental procedure [239].

If δL is nonzero, the measured phase function, $\theta_{\text{exp}}(\omega)$, as a function of angular frequency, is given by

$$\theta_{\text{exp}}(\omega) = \theta(\omega) + \alpha\omega, \tag{10.35}$$

where $\theta(\omega)$ is the true phase, $\alpha = c^{-1}\delta L$ is a phase error constant, and c is the speed of light in vacuum. The main idea in using the MEM fitting of $r(\omega)$ for

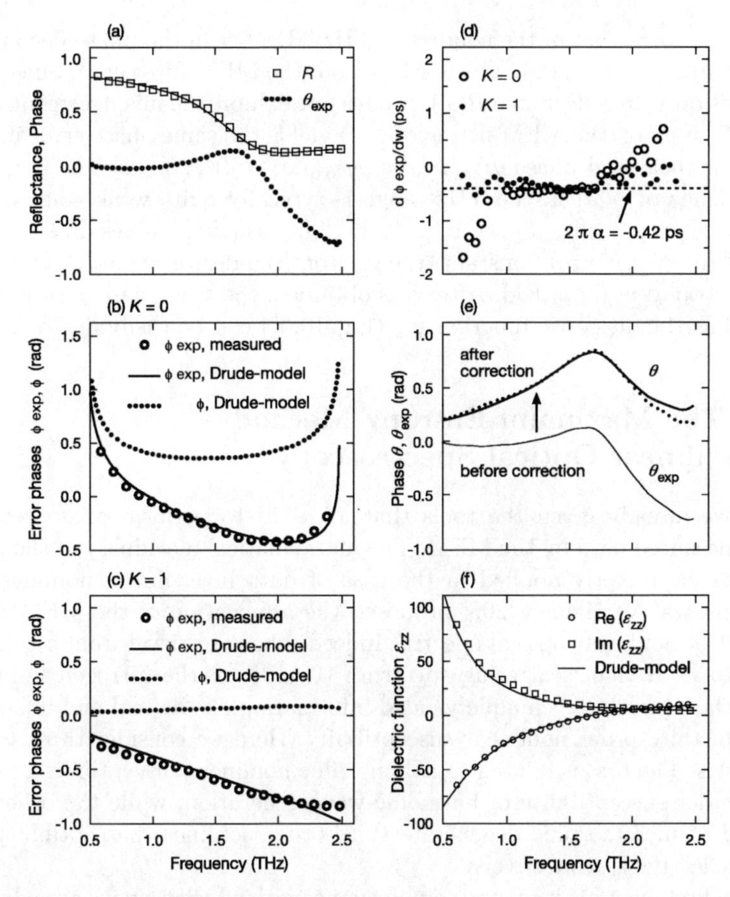

Fig. 10.6. (**a**) Measured THz reflectance (*open squares*) and the phase (*dotted line*) with the corresponding Drude model reflectance (*solid line*). (**b**) The measured ME error phase (*open circles*) and the corresponding simulated curves with (*solid line*) and without (*dotted line*) the phase error. The squeezing parameter is set to $K = 0$. In the simulation, the phase error constant is set to $\alpha = 0.42$. (**c**) The same as in (**b**), but with $K = 1$. (**d**) Experimental derivatives $\mathrm{d}\phi_{\exp}(\omega; K)/\mathrm{d}\omega$ with $K = 0$ (*open circles*) and $K = 1$ (*dots*) revealing the phase error constant α and, thereby, (**e**) the phase correction. *Solid lines* are uncorrected and corrected phase measurements. The *dotted line* shows the phase obtained by the Kramers-Kronig analysis, where the reflectance was extrapolated with the Drude model. (**f**) Determined complex dielectric function, ε_{zz} (*dots*), and the corresponding calculation by the Drude model (*solid lines*). Reproduced from [239]

the phase correction in the context of THz-TDS lies in the phase decomposition, $\theta(\omega) = \phi(\omega) + \psi(\omega)$, which arises from the MEM fit of $r(\omega)$. Since $\psi(\omega)$ depends only on reflectance $R(\omega)$, it does not exhibit the misplacement phase error. Therefore, the MEM fitting of $r(\omega)$ yields the same phase error in $\phi(\omega)$ as in the measured phase $\theta_{\text{exp}}(\omega)$, i.e., $\phi_{\text{exp}}(\omega) = \phi(\omega) + \alpha\omega$. The frequency dependence of $\phi(\omega)$ in the THz range is typically quite weak compared to the phase error term $\alpha\omega$ (see Fig. 10.6c). This usually enables direct extraction of the phase error constant α, e.g., from the derivative $d\phi_{\text{exp}}(\omega)/d\omega$, as demonstrated in Fig. 10.6d. After α is obtained, the true phase φ (Fig. 10.6e) as well as the dielectric function ε_{zz} (Fig. 10.6f) can be easily derived.

10.3 The Maximum Entropy Method in Nonlinear Optical Spectroscopy

We have already given the tools that allow us to perform phase retrieval from measured data by the MEM. The mathematical machinery of the MEM can also be directly applied in the case of data inversion of nonlinear optical spectra. Vartiainen [206] proposed the application of the MEM in the context of nonlinear optical spectra. Indeed, phase retrieval from a coherent anti-Stokes Raman scattering spectrum (CARS) of the nitrogen Q-branch using the squeezing technique yielded information on the real and imaginary parts of third-order nonlinear susceptibility. Here we consider three typical examples. The first is related to holomorphic nonlinear susceptibility, i.e. the third-order susceptibility of harmonic-wave generation, while the others are related to meromorphic degenerate third-order nonlinear susceptibility and total reflectivity, respectively.

The first example is related to polysilane, with which we have already been dealing in the context of K-K relations. Here we consider only the comparison between the MEM and K-K methods. More detailed results can be found in [206]. In Fig. 10.7a,b, we show the real and imaginary parts of the third-order harmonic susceptibility of polysilane retrieved by two alternative methods. The MEM curves were computed by estimating the error phase with first-order ($L = 1$) and third-order ($L = 3$) polynomials. We can observe that the two MEM estimates give, practically speaking, identical curves. Furthermore, if we compare K-K and MEM, we observe that the curves resemble each other very much. However, K-K analysis could not reproduce the two-photon resonance peak at 2.1 eV. In both analyses, the phase data were required for two wavelengths. The significant difference was that in the MEM, the anchor points were inside the measured range, whereas in the K-K analysis, they were outside the measured range.

The second example is a simulation related to the effective degenerate third-order nonlinear susceptibility of a MG two-phase nanocomposite. In the simulation, the host material is assumed to respond linearly to intense light, whereas the inclusions are assumed to have a nonlinear response. In

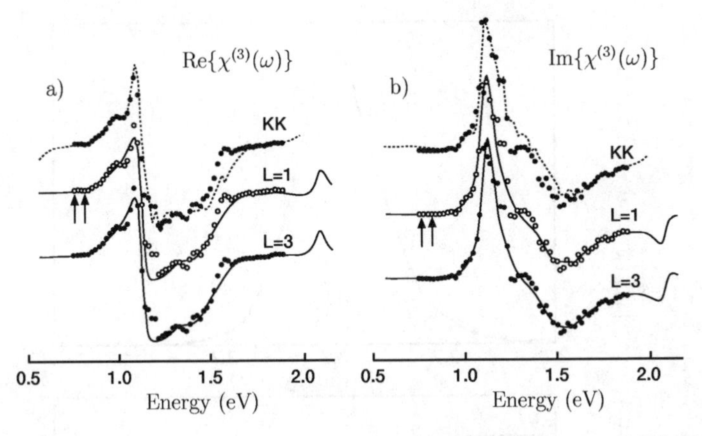

Fig. 10.7. Experimental values (*dots* and *open circles*) of (**a**) real and (**b**) imaginary parts of susceptibility $\chi^{(3)}(\omega;\omega,\omega,-\omega)$ of polydihexylsilane, and the corresponding curves obtained by Kramers-Kronig analysis (*dotted lines*) and by the MEM procedure (*solid lines*). The additional information used for MEM estimates with L = 1 was the two phase values indicated by the *arrows*. Reproduced from [206]

such a case the effective nonlinear susceptibility can be found with the aid of (5.37)–(5.41) as follows:

$$\chi^{(3)}_{\mathrm{eff}}(\omega;\omega,\omega,-\omega) = f\left|\frac{\varepsilon_{\mathrm{eff}}(\omega)+2\varepsilon_{\mathrm{h}}(\omega)}{\varepsilon_{\mathrm{i}}(\omega)+2\varepsilon_{\mathrm{h}}(\omega)}\right|^2\left[\frac{\varepsilon_{\mathrm{eff}}(\omega)+2\varepsilon_{\mathrm{h}}(\omega)}{\varepsilon_{\mathrm{i}}(\omega)+2\varepsilon_{\mathrm{h}}(\omega)}\right]^2 \times \chi^{(3)}_{\mathrm{i}}(\omega;\omega,\omega,-\omega), \tag{10.36}$$

where we assume that

$$\chi^{(3)}_{\mathrm{i}}(\omega;\omega,\omega,-\omega) = \frac{B}{\left|\omega_0^2-\omega^2-\mathrm{i}\gamma\omega\right|^2\left(\omega_0^2-\omega^2-\mathrm{i}\gamma\omega\right)^2}, \tag{10.37}$$

and B is a constant. We next resolve the real and imaginary parts from the modulus of the effective degenerate nonlinear susceptibility by the MEM. In the case of the four chosen anchor points, the results are shown in Fig. 10.8. K-K relations cannot be used in a problem like this since they result in erroneous data inversions (see [15]).

The third example is devoted to meromorphic total reflectance. In other words, once again we have meromorphic third-order nonlinear susceptibility, but this time, we assume that the system involves a two-phase Bruggeman liquid. This means that we consider the expression (5.42) and assume that only the nanoparticles are optically nonlinear with a susceptibility given by (10.37), whereas the liquid matrix is optically linear. As usual, the optical properties of the Bruggeman as well as of other liquids can be searched through a probe window. However, in the case of intense light obtained from a tunable dye laser, the reflected light has linear and nonlinear contributions.

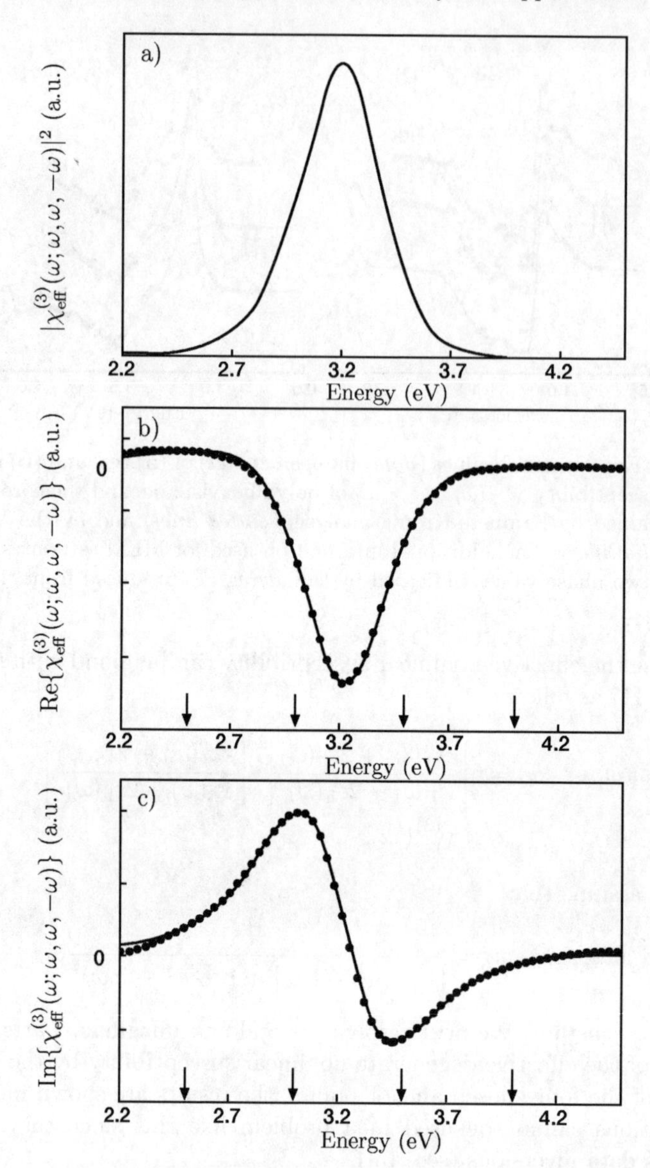

Fig. 10.8. (**a**) Squared modulus of the effective nonlinear susceptibility of Maxwell Garnett material. Real (**b**) and imaginary (**c**) parts obtained by the MEM procedure, where additional information about the phase is assumed to be known for four equispaced frequencies (*arrows*). Reproduced from [15]

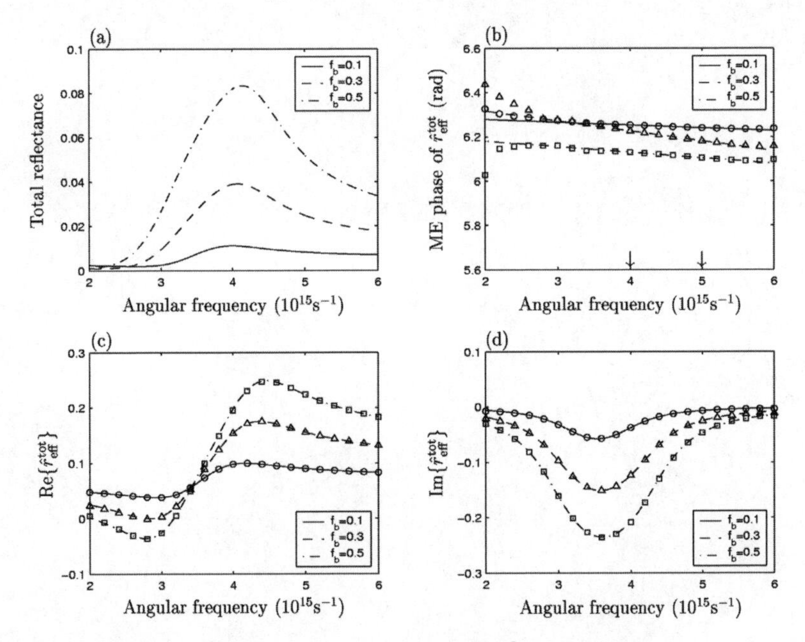

Fig. 10.9. (**a**) Effective total reflectance $R = |\hat{r}_{\text{eff}}^{\text{tot}}|^2$; (**b**) maximum entropy phase (*line curves*) and corresponding linear approximate (*dots*); (**c**) real, and (**d**) imaginary part of the total reflectivity $\hat{r}_{\text{eff}}^{\text{tot}}$ (*line curves*), and the corresponding MEM estimates (*dots*). The *arrows* in (**b**) indicate the frequencies at which the phase values were to be known a priori. A=10, $\omega_0 = 3 \times 10^{15}$ s^{-1}, $B = 1$, $\varepsilon_{\text{p}} = 2.25$, and $I = |E|^2 = 10$ (which corresponds to a true field strength $E = 10^9$ V/m). *Solid lines*: $f_{\text{b}} = 0.1$. *Dashed lines*: $f_{\text{b}} = 0.3$. *Dash-dotted lines*: $f_{\text{b}} = 0.5$. Reproduced from [71]

Therefore we have to consider the total reflectance since both the linear and nonlinear contributions have the same angular frequency and propagate spatially along the same path to the detector system. Now in the case of normal light incidence, in internal reflection, the total reflectance is as follows:

$$R(\omega) = \left| \frac{1 - \sqrt{\varepsilon_{\text{eff}}^{\text{tot}}(\omega)/\varepsilon_{\text{p}}(\omega)}}{1 + \sqrt{\varepsilon_{\text{eff}}^{\text{tot}}(\omega)/\varepsilon_{\text{p}}(\omega)}} \right|, \tag{10.38}$$

where ε_{p} is the dielectric function of the probe window and

$$\varepsilon_{\text{eff}}^{\text{tot}} = \varepsilon_{\text{eff}}^{(1)} + \chi_{\text{eff}}^{(3)} I, \tag{10.39}$$

where I is the intensity of light. We show in Fig. 10.9 the results of the MEM analysis where the real and imaginary parts of total reflectance have been resolved for the cases of two anchor points and three different fill fractions of the nanofeatures.

11 Conclusions

The purpose of this book has been to highlight the relevance of the general integral properties of susceptibility functions for the inspection of the linear and nonlinear optical properties of materials. The conceptual foundation of the general integral properties set forth is the principle of causality in light–matter interaction. The bridge between the causality of the physical system and the holomorphic properties of the susceptibilities analyzed, which determine the possibility of writing the integral relations, is Titchmarsch's theorem in the case of linear optics and Scandolo's theorem in the nonlinear case. We have shown that, in all generality, such general properties only depend on the expectation value of suitable operators in the ground state of the electronic density matrix densities, with appropriate modifications to account for local field effects and for inhomogeneous media. Moreover, since we have adopted a rather general quantum mechanical framework, these results are derived for any physical system.

In the case of linear optics, we have reviewed in great detail the theoretical derivation and the actual experimental use of Kramers-Kronig relations for susceptibility as well as for the relevant optical constants, reflectance and index of refraction. We have also shown how, using the superconvergence theorem on such dispersion relations and comparing the results with the correct asymptotic behavior obtained from quantum response theory, it is possible to derive sum rules, which constitute stringent constraints having profound relevance in the interpretation of both model-generated and experimental data. We have also shown how these linear integral properties are changed when conductors are considered.

In the nonlinear case, the susceptibility functions that have a holomorphic character in the half complex plane of the relevant frequency variable are shown to obey a number of dispersion relations between their real and imaginary parts. The important cases of pump-and-probe systems and of harmonic-generation processes are analyzed in detail and new Kramers-Kronig relations are given for susceptibilities and their powers. In general, the number of Kramers-Kronig type relations depends on the asymptotic behavior of nonlinear susceptibility for large values of the relevant frequency variable. Subtractive Kramers-Kronig relations are derived to improve the retrieval process.

Using the superconvergence theorem on such dispersion relations and comparing the results with the correct asymptotic behavior obtained from quantum response theory, new sum rules are obtained for nonlinear susceptibilities to any order. The sum rules can be used to check any approximate theory of nonlinear phenomena, because they are necessary constraints which have to be obeyed. In the case of harmonic-generation susceptibilities, such sum rules have the common features that all even moments of the real part and all odd moments of the imaginary parts are null, except for the highest converging odd moment of the imaginary part of the susceptibility, which gives a result dependent on the expectation value in the ground state of appropriate derivatives of the potential. This gives a qualitative measure of the optical nonlinearity of the material.

Kramers-Kronig type relations are shown to be useful for data inversions, required because in most materials only one optical function can be measured and the range of frequency attainable is rather narrow. Specific examples of the application of conventional, generalized, and subtractive Kramers-Kronig relations as well as of verification of sum rules are given for experimental data on third-harmonic-generation in polymers. In a relevant example of second-harmonic-generation, it is shown that sum rules allow of better understanding of Miller's rule and provide tools for explaining critical experimental parameters.

Therefore, by analyzing actual experimental data, we have shown that, far from being somewhat cumbersome and fictitious theoretical findings, these integral properties have a huge potential for providing new tools for profound, self-consistent analysis of a new generation of experimental and model-generated data concerned with nonlinear optical processes. Integral relations form the best language for use in speaking about frequency-dependent nonlinear optical phenomena. We believe that the role of dispersion theory and sum rules will be of increasing importance in the expanding field of nonlinear optics.

In the case of meromorphic nonlinear susceptibility, conventional K-K relations have to be modified. By taking into account the poles that are located in the upper half plane, we derive modified K-K relations and sum rules for meromorphic nonlinear susceptibilities. Furthermore, by applying a partial fraction to the poles, we observe new practical sum rules for meromorphic nonlinear susceptibilities such as at describing degenerate four-wave mixing.

Unfortunately, in linear reflectance and in most measurements of nonlinear susceptibility, only the intensity data on the measured quantity are available. Thus, a phase retrieval problem arises. Both multiply-subtractive K-K relations and the MEM procedure are alternative methods that can be applied both for linear reflectance and for the modulus of nonlinear susceptibility. Furthermore, they can be exploited for finite frequency data analysis. Unfortunately, for meromorphic nonlinear susceptibilities, the conventional K-K analysis is not valid. However, the MEM can be exploited for phase retrieval both for linear and nonlinear susceptibilities, including phase retrieval from nanostructures.

Given the very abstract nature of the physical problem we have examined, at least in theoretical terms, and given that we set out with the aim of providing a unifying picture, we have followed a wholly deductive argumentation, progressing from linear to nonlinear, from the general to the specific, and from the theoretical to the experimental. Within a single perspective, we have framed many results that have otherwise been sparsely presented in the literature, and we have proposed new theoretical tools for investigation. We hope that this effort will form a baseline for future theoretical and experimental research in wide spectral range nonlinear optical investigation.

A MATLAB® Programs for Data Analysis

In this appendix, we present some basic programs written for the MATLAB® environment for the analysis of the data. These programs can be easily customized by the expert user, but they nevertheless constitute useful data analysis tools also in the present form.

The first two programs deal with the computation of K-K relations. The third program can be used to obtain self-consistent (in terms of K-K relations) estimates of the real and the imaginary parts of susceptibility when first-guess estimates are used as input. It is particularly suitable when the first-guess estimates of the real and imaginary parts of susceptibility have been independently obtained, e.g. by direct measurements. The fourth and the fifth program deal with the computation of SSKK relations.

In order to take advantage of this set of programs, and considering that the first two programs are called by the last three programs, it is strongly advised to save them in the same directory with the following names:

- Program 1: kkimbook.m;
- Program 2: kkrebook.m;
- Program 3: selfconsbook.m;
- Program 4: sskkimbook.m;
- Program 5: sskkrebook.m.

These programs require that the spectral data given as input have constant frequency spacing. Simple interpolation schemes can in most cases efficiently rearrange diversely spaced data to this form. These programs have been tested on the MATLAB® versions 6.x both for Linux/Unix® and Microsoft Windows® environments.

A.1 Program 1: Estimation of the Imaginary Part via Kramers-Kronig Relations

```
function imchi=kkimbook(omega,rechi,alpha)
%The program inputs are 1) omega, vector of the frequency
%(or energy) components, 2) rechi, vector of the real part
%of the susceptibility under examination, 3) alpha, value
```

```
%of the moment considered. The two vectors 1) and 2)
%must have the same length. The output is the estimate
%of the imaginary part as obtained with K-K relations.
%In order to use this program, save the whole text contained
%in this section in a file and name it kkimbook.m

if size(omega,1)>size(omega,2);
    omega=omega';
end; if size(rechi,1)>size(rechi,2);
    rechi=rechi';
end;
%Here the program rearranges the two vectors so that,
%whichever their initial shape, they become row vectors.

g=size(omega,2);
%Size of the vectors.%

imchi=zeros(size(rechi));
%The output is initialized.

a=zeros(size(rechi));
b=zeros(size(rechi));
%Two vectors for intermediate calculations are initialized

deltaomega=omega(2)-omega(1);
%Here we compute the frequency (or energy) interval

j=1;
beta1=0;
for k=2:g;
    b(1)=beta1+rechi(k)*omega(k)^(2*alpha)/(omega(k)^2-omega(1)^2);
    beta1=b(1);
end;
imchi(1)=-2/pi*deltaomega*b(1)*omega(1)^(1-2*alpha);
%First element of the output: the principal part integration
%is computed by excluding the first element of the input

j=g;
alpha1=0;
for k=1:g-1;
    a(g)=alpha1+rechi(k)*omega(k)^(2*alpha)/(omega(k)^2-omega(g)^2);
    alpha1=a(g);
end;
imchi(g)=-2/pi*deltaomega*a(g)*omega(g)^(1-2*alpha);
%Last element of the output: the principal part integration
%is computed by excluding the last element of the input.

for j=2:g-1; ;
```

```
%Loop on the inner components of the output vector.
    alpha1=0;
    beta1=0;
    for k=1:j-1;
        a(j)=alpha1+rechi(k)*omega(k)^(2*alpha)/...
                                    (omega(k)^2-omega(j)^2);
        alpha1=a(j);
    end;
    for k=j+1:g;
        b(j)=beta1+rechi(k)*omega(k)^(2*alpha)/...
                                    (omega(k)^2-omega(j)^2);
        beta1=b(j);
    end;
    imchi(j)=-2/pi*deltaomega*(a(j)+b(j))*omega(j)^(1-2*alpha);
end;
%Last element of the output: the principal part integration
%is computed by excluding the last element of the input
```

A.2 Program 2: Estimation of the Real via Kramers-Kronig Relations

```
function rechi=kkrebook(omega,imchi,alpha)
%The program inputs are 1) omega, vector of the frequency
%(or energy) components, 2) imchi, vector of the imaginary
%part of the susceptibility under examination, and 3) alpha,
%the value of the moment considered. The two vectors
%1) and 2) must have the same length.
%The output is the estimate of the real part as obtained
%with K-K relations.
%In order to use this program, save the whole text contained
%in this section in a file and name it kkrebook.m

if size(omega,1)>size(omega,2);
    omega=omega';
end; if size(imchi,1)>size(imchi,2);
    imchi=imchi';
end;
%Here the program rearranges the two vectors so that,
%whichever their initial shape, they become row vectors.

g=size(omega,2);
%Size of the vectors.%

rechi=zeros(size(imchi));
%The output is initialized.

a=zeros(size(imchi));
```

```
b=zeros(size(imchi));
%Two vectors for intermediate calculations are initialized

deltaomega=omega(2)-omega(1);
%Here we compute the frequency (or energy) interval

j=1;
beta1=0;
for k=2:g;
    b(1)=beta1+imchi(k)*omega(k)^(2*alpha+1)/...
                                          (omega(k)^2-omega(1)^2);
    beta1=b(1);
end;
rechi(1)=2/pi*deltaomega*b(1)*omega(1)^(-2*alpha);
%First element of the output: the principal part integration
%is computed by excluding the first element of the input

j=g;
alpha1=0;
for k=1:g-1;
    a(g)=alpha1+imchi(k)*omega(k)^(2*alpha+1)/...
                                          (omega(k)^2-omega(g)^2);
    alpha1=a(g);
end;
rechi(g)=2/pi*deltaomega*a(g)*omega(g)^(-2*alpha);
%Last element of the output: the principal part integration
%is computed by excluding the last element of the input

for j=2:g-1; ;
%Loop on the inner components of the output vector.
    alpha1=0;
    beta1=0;
    for k=1:j-1;
        a(j)=alpha1+imchi(k)*omega(k)^(2*alpha+1)/...
                                          (omega(k)^2-omega(j)^2);
        alpha1=a(j);
    end;
    for k=j+1:g;
        b(j)=beta1+imchi(k)*omega(k)^(2*alpha+1)/...
                                          (omega(k)^2-omega(j)^2);
        beta1=b(j);
    end;
    rechi(j)=2/pi*deltaomega*(a(j)+b(j))*omega(j)^(-2*alpha);
end;
%Last element of the output: the principal part integration
%is computed by excluding the last element of the input
```

A.3 Program 3: Self-Consistent Estimate of the Real and Imaginary Parts of Susceptibility

```
function [refin,imfin]=selfconsbook(omega,rechi,imchi,N,mu)
%The program inputs are 1) omega, vector of the frequency (or
%energy) vector, 2) rechi, vector of the first-guess real part
%of the susceptibility under examination 3) imchi, vector of
%the first-guess imaginary part of the susceptibility under
%examination, 4) N, number of iterations, and 5) mu, weight factor,
%which must be between 0 and 1 (0.5 is usually a good choice).
%mu determines the weight we want to give to the first-guess
%estimates in the self-consistent procedure. The three vectors
%1), 2) and 3) must have the same length.
%The output consists of the self-consistent estimates of
%the real and of the imaginary part as obtained by
%combining recursively first-guess estimates and outputs
%of K-K relations.
%In order to use this program, save the whole text contained
%in this section in a file and name it selfconsbook.m

if size(omega,1)>size(omega,2);
    omega=omega';
end;
if size(rechi,1)>size(rechi,2);
    rechi=rechi';
end;
if size(imchi,1)>size(imchi,2);
    imchi=imchi';
end;
%Here the program rearranges the three vectors so that,
%whichever their initial shape, they become row vectors.

comodo1=rechi;
comodo2=imchi;
%Here the program defines two intermediate variables.

for j=1:N;
    comodo1=kkrebook(omega,comodo2,0);
    comodo1=(mu*rechi+(1-mu)*comodo1);
    comodo2=kkimbook(omega,comodo1,0);
    comodo2=(mu*imchi+(1-mu)*comodo2);
end;
%At each step the program computes the best estimate
%of the real and imaginary part by combining application
%of K-K relations with actual measurements.
refin=comodo1;
imfin=comodo2;
```

A.4 Program 4: Estimation of the Imaginary Part via Singly Subtractive Kramers-Kronig Relations

```
function imchi=sskkimbook(omega,rechi,omega1,imchi1,alpha)
%The program inputs are 1) omega, vector of the frequency
%(or energy) components, 2) rechi, vector of the real
%part of the susceptibility under examination, 3) omega1,
%anchor point, 4) imchi1, value of the imaginary part at
%the anchor point, 5) alpha, value of the moment considered.
%The two vectors 1) and 2) must have the same length.
%The output is the estimate of the
%imaginary part as obtained by using SSKK relations.
%In order to use this program, save the whole text contained
%in this section in a file and name it sskkimbook.m

if size(omega,1)>size(omega,2);
   omega=omega';
end; if size(rechi,1)>size(rechi,2);
   rechi=rechi';
end;
%Here the program rearranges the two vectors so that,
%whichever their initial shape, they become row vectors.

g=size(omega,2);
%Size of the vectors.%

k=0;
for j=1:g;
   if omega(j)==omega1;
      k=j;
   end;
end;
%Determination of the anchor point.

imchi=kkimbook(omega,rechi,alpha);
%Application of K-K relations

imchi=imchi+omega1^(2*alpha-1)*omega.^(1-2*alpha)*(imchi1-imchi(k));
%The subtracted relation upgrades the estimate obtained
%with K-K relations.
```

A.5 Program 5: Estimation of the Real Part via Singly Subtractive Kramers-Kronig Relations

```
function rechi=sskkrebook(omega,imchi,omega1,rechi1,alpha)
%The program inputs are 1) omega, vector of the
% frequency (or energy) components, 2) imchi, vector of
%the imaginary part of the susceptibility
%under examination, 3) omega1, anchor point, 4) rechi1,
%value of the real part at the anchor point, 5) alpha,
%value of the moment considered.
%The two vectors 1) and 2) must have the same length.
%The output is the estimate of the
%real part as obtained by using SSKK relations.
%In order to use this program, save the whole text contained
%in this section in a file and name it sskkrebook.m

if size(omega,1)>size(omega,2);
    omega=omega';
end; if size(imchi,1)>size(imchi,2);
    rechi=rechi';
end;
%Here the program rearranges the two vectors so that,
%whichever their initial shape, they become row vectors.

g=size(omega,2);
%Size of the vectors.%

k=0; for j=1:g;
    if omega(j)==omega1;
        k=j;
    end;
end;
%Determination of the anchor point.

rechi=kkrebook(omega,imchi,alpha);
%Application of K-K relations

rechi=rechi+omega1^(2*alpha)*omega.^(-2*alpha)*(rechi1-rechi(k));
%The subtracted relation upgrades the estimate obtained
%with K-K relations.
```

References

1. J. S. Toll, "Causality and the dispersion relation: Logical foundations," *Phys. Rev.* **104**, 1760–1770 (1956).
2. H. M. Nussenzveig, *Causality and Dispersion Relations* (Academic, New York, 1972).
3. P. W. Milonni, "Controlling the speed of light pulses," *J. Phys. B: At. Mol. Opt. Phys.* **35**, R31–R56 (2002).
4. H. A. Kramers, "La diffusion de la lumiére par les atomes," in *Atti del Congresso Internazionale dei Fisici*, Vol. 2 (Zanichelli, Bologna, 1927), pp. 545–557.
5. R. de L. Kronig, "On the theory of dispersion of x-rays," *J. Opt. Soc. Am.* **12**, 547–557 (1926).
6. W. Thomas, "Über die zahl der dispersionselektronen, die einem stationären zustande zugeordnet sind," *Naturwiss.* **28**, 627 (1925).
7. F. Reiche and W. Thomas, "Über die zahl der dispersionselektronen, die einem stationären zustand zugeordnet sind," *Z. Phys.* **34**, 510–525 (1925).
8. W. Kühn, "Über die gesamtstärke der von einem zustande ausgehenden absorptionslinien," *Z. Phys.* **33**, 408–412 (1925).
9. M. Altarelli, D. L. Dexter, H. M. Nussenzveig, and D. Y. Smith, "Superconvergence and sum rules for the optical constants," *Phys. Rev. B* **6**, 4502–4509 (1972).
10. D. L. Greenaway and G. Harbeke, *Optical Properties and Band Structure of Semiconductors* (Pergamon, Oxford, 1968).
11. L. D. Landau and E. M. Lifshitz, *Statistical Physics: Part I* (Mir Editions, Moscow, 1972).
12. F. Bassani and M. Altarelli, "Interaction of radiation with condensed matter," in *Handbook on Synchrotron Radiation*, E. E. Koch, ed. (North Holland, Amsterdam, 1983), pp. 465–597.
13. D. E. Aspnes, "The accurate determination of optical properties by ellipsometry," in *Handbook of Optical Constants of Solids*, E. D. Palik, ed. (Academic, New York, 1985), pp. 89–112.
14. M. Cardona and P. Y. Yu, *Fundamentals of Semiconductors* (Springer, Berlin, 1996).
15. K.-E. Peiponen, E. M. Vartiainen, and T. Asakura, *Dispersion, Complex Analysis and Optical Spectroscopy* (Springer, Heidelberg, 1999).
16. R. P. Feynman, "Space-time approach to quantum electrodynamics," *Phys. Rev.* **76**, 769–789 (1949).
17. N. Bloembergen, *Nonlinear Optics* (Benjamin, New York, 1965).
18. P. N. Butcher and D. Cotter, *Elements of Nonlinear Optics* (Cambridge University Press, Cambridge, 1990).

19. R. W. Boyd, *Nonlinear Optics* (Academic Press, New York, 1992).
20. K.-E. Peiponen, "Sum rules for the nonlinear susceptibilities in the case of sum frequency generation," *Phys. Rev. B* **35**, 4116–4117 (1987).
21. K.-E. Peiponen, "On the derivation of sum rules for physical quantities and their applications for linear and non-linear optical constants," *J. Phys. C.: Solid State Phys.* **20**, 2785–2788 (1987).
22. K.-E. Peiponen, "Nonlinear susceptibilities as a function of several complex angular-frequency variables," *Phys. Rev. B* **37**, 6463–6467 (1988).
23. F. Bassani and S. Scandolo, "Dispersion relations and sum rules in nonlinear optics," *Phys. Rev. B* **44**, 8446–8453 (1991).
24. F. Bassani and S. Scandolo, "Sum rules for nonlinear optical susceptibilities," *Phys. Stat. Sol. (b)* **173**, 263–270 (1992).
25. S. Scandolo and F. Bassani, "Nonlinear sum rules: The three level and the anharmonic-oscillator models," *Phys. Rev. B* **45**, 13257–13261 (1992).
26. D. C. Hutchings, M. Sheik-Bahae, D. J. Hagan, and E. W. van Stryland, "Kramers-Krönig relations in nonlinear optics," *Optical and Quantum Electronics* **24**, 1–30 (1992).
27. Y. H. Lee, A. Chavez-Pirson, S. W. Koch, H. M. Gibbs, S. H. Park, J. Morhange, A. Jeffrey, N. Peyghambarian, L. B. A. C. Gossard, and W. Wiegmann, "Room-temperature optical nonlinearities in GaAs," *Phys. Rev. Lett.* **57**, 2446–2449 (1986).
28. D. Guo, S. Mazumdar, G. I. Stegeman, M. Cha, D. Neher, S. Aramaki, W. Torruellas, and R. Zanoni, "Nonlinear optics of linear conjugated polymers," *Mater. Res. Soc. Symp. Proc.* **247**, 151–162 (1992).
29. H. Kishida, T. Hasegawa, Y. Iwasa, T. Koda, and Y. Tokura, "Dispersion relation in the third-order electric susceptibility for polysilane film," *Phys. Rev. Lett.* **70**, 3724–3727 (1993).
30. F. S. Cataliotti, C. Fort, T. W. Hänsch, M. Inguscio, and M. Prevedelli, "Electromagnetically induced transparency in cold free atoms: Test of a sum rule for nonlinear optics," *Phys. Rev. A* **56**, 2221–2224 (1997).
31. V. Lucarini, J. J. Saarinen, and K.-E. Peiponen, "Multiply subtractive Kramers–Kronig relations for arbitrary-order harmonic generation susceptibilities," *Opt. Commun.* **218**, 409–414 (2003).
32. V. Lucarini and K.-E. Peiponen, "Verification of generalized Kramers-Kronig relations and sum rules on experimental data of third harmonic generation susceptibility on polymer," *J. Phys. Chem.* **119**, 620–627 (2003).
33. V. Lucarini, J. J. Saarinen, and K.-E. Peiponen, "Multiply subtractive generalized Kramers–Kronig relations: Application on third-harmonic generation susceptibility on polysilane," *J. Chem. Phys.* **119**, 11095–11098 (2003).
34. V. Lucarini, F. Bassani, K.-E. Peiponen, and J. J. Saarinen, "Dispersion relations in linear and nonlinear optics," *Riv. Nuovo Cim.* **26**, 1–120 (2003).
35. J. D. Jackson, *Classical Electrodynamics*, 3rd ed. (Wiley, New York, 1998).
36. B. I. Bleaney and B. Bleaney, *Electricity and Magnetism*, 3rd ed. (Oxford University Press, London, 1976).
37. M. Born and E. Wolf, *Principles of Optics*, 7th ed. (Cambridge University Press, Cambridge, 1999).
38. L. D. Landau, E. M. Lifshitz, and L. P. Pitaevskii, *Electrodynamics of Continuous Media*, 2nd ed. (Pergamon, Oxford, 1984).
39. G. B. Arfken and H. J. Weber, *Mathematical Methods for Physicists*, 4th ed. (Academic Press, San Diego, 1995).

40. L. D. Landau and E. M. Lifshitz, *Mechanics* (Pergamon, Oxford, 1976).
41. M. Goppert-Mayer, "Über elementarakte mit zwei quantens prungen," *Ann. Phys.* **40**, 273–295 (1931).
42. F. Bassani, J. J. Forney, and A. Quattropani, "Choice of gauge in two-photon transitions: 1s-2s transition in atomic hydrogen," *Phys. Rev. Lett.* **39**, 1070–1073 (1977).
43. A. Quattropani, F. Bassani, and S. Carillo, "Two photon transition to excited states in atomic hydrogen," *Phys. Rev. A* **25**, 3079–3089 (1982).
44. J. van Kranendonk and J. E. Sipe, "Foundations of the macroscopic electromagnetic theory of dielectric media," in *Progress in Optics*, Vol. XV, E. Wolf, ed. (Elsevier, Amsterdam, 1977).
45. J. J. Maki, M. S. Malcuit, J. E. Sipe, and R. W. Boyd, "Linear and nonlinear optical measurements of the Lorentz local field," *Phys. Rev. Lett.* **67**, 972–975 (1991).
46. D. E. Aspnes, "Local-field effects and effective-medium theory: A microscopic perspective," *Am. J. Phys.* **50**, 704–709 (1982).
47. X. C. Zeng, D. J. Bergman, P. M. Hui, and D. Stroud, "Effective-medium theory for weakly nonlinear composites," *Phys. Rev. B* **38**, 10970–10973 (1988).
48. R. W. Boyd and J. E. Sipe, "Nonlinear optical susceptibilities of layered composite materials," *J. Opt. Soc. Am. B* **11**, 297–303 (1994).
49. B. N. A. Lagendijk, B. A. van Tiggelen, and P. de Vries, "Microscopic approach to the Lorentz cavity in dielectrics," *Phys. Rev. Lett.* **79**, 657–660 (1997).
50. G. Grosso and G. P. Parravicini, *Solid State Physics* (Academic Press, San Diego, 2000).
51. R. L. Sutherland, *Handbook of Nonlinear Optics* (Marcel Dekker, New York, 1996).
52. J. C. Maxwell Garnett, "Colours in metal glasses and in metallic films," *Philos. Trans. R. Soc.* **203**, 385–420 (1904).
53. J. C. Maxwell Garnett, "Colours in metallic glasses, in metallic films, and in metallic solution," *Philos. Trans. R. Soc.* **205**, 237–288 (1906).
54. J. I. Gittleman and B. Abeles, "Comparison of the effective medium and the Maxwell-Garnett predictions for the dielectric constants of granular metals," *Phys. Rev. B* **15**, 3273–3275 (1977).
55. D. A. G. Bruggeman, "Berechnung verschiedener physikalischer konstanten von heterogenen substanzen. I. Dielektizitätskonstanten und leitfähigkeiten der mischkörper aus isotropen substanzen," *Ann. Phys. (Leipzig)* **24**, 636–679 (1935).
56. R. Landauer, "The electrical resistance of binary metallic mixtures," *J. Appl. Phys.* **23**, 779–784 (1952).
57. P. Chýlek and G. Videen, "Scattering by a composite sphere and effective medium approximations," *Opt. Commun.* **146**, 15–20 (1998).
58. R. Ruppin, "Evaluation of extented Maxwell-Garnett theories," *Opt. Commun.* **182**, 273–279 (2000).
59. R. W. Boyd, R. J. Gehr, G. L. Fischer, and J. E. Sipe, "Nonlinear optical properties of nanocomposite materials," *Pure Appl. Opt.* **5**, 505–512 (1996).
60. D. Prot, D. B. Stroud, J. Lafait, N. Pinçon, B. Palpant, and S. Debrus, "Local electric field enhancements and large third-order optical nonlinearity in nanocomposite materials," *Pure Appl. Opt.* **4**, S99–S102 (2002).
61. D. Dalacu and L. Martinu, "Spectroellipsometric characterization of plasma-deposited Au/SiO_2 nanocomposite films," *J. Appl. Phys.* **87**, 228–235 (2000).

62. K.-E. Peiponen, M. O. A. Mäkinen, J. J. Saarinen, and T. Asakura, "Dispersion theory of liquids containing optically linear and nonlinear Maxwell Garnett nanoparticles," *Opt. Rev.* **8**, 9–17 (2001).
63. N. J. Harrick, *Internal Reflection Spectroscopy* (Harrick Scientific Corporation, New York, 1979).
64. R. J. Gehr, G. L. Fischer, and R. W. Boyd, "Nonlinear-optical response of porous-glass-based composite materials," *J. Opt. Soc. Am. B* **14**, 2310–2314 (1997).
65. W. Schützer and J. Tiomno, "On the connection of the scattering and derivative matrices with causality," *Phys. Rev.* **83**, 249–251 (1951).
66. N. G. van Kampen, "S-matrix and causality condition. I. Maxwell field," *Phys. Rev.* **89**, 1072–1079 (1953).
67. N. G. van Kampen, "S-matrix and causality condition. II. Nonrelativistic particles," *Phys. Rev.* **91**, 1267–1276 (1953).
68. E. Shiles, T. Sasaki, M. Inokuti, and D. Y. Smith, "Self-consistency and sum-rule tests in the Kramers-Kronig analysis of optical data: Applications to aluminum," *Phys. Rev. B* **22**, 1612–1628 (1980).
69. B. J. Kowalski, A. Sarem, and B. A. Orowski, "Optical parameters of $Cd_{1-x}Fe_xSe$ and $Cd_{1-x}Fe_xTe$ by means of Kramers-Kronig analysis of reflectivity data," *Phys. Rev. B* **42**, 5159–5165 (1990).
70. D. Miller and P. L. Richards, "Use of Kramers-Kronig relations to extract the conductivity of high-Tc superconductors from optical data," *Phys. Rev. B* **47**, 12308–12311 (1993).
71. K.-E. Peiponen, E. M. Vartiainen, J. J. Saarinen, and M. O. A. Mäkinen, "The dispersion theory of optically linear and nonlinear Bruggeman liquids," *Opt. Commun.* **205**, 17–24 (2002).
72. K.-E. Peiponen and T. Asakura, "Dispersion theory for two-phase layered-geometry nanocomposites," *Opt. Rev.* **6**, 410–414 (1999).
73. D. Stroud, "Percolation effects and sum rules in the optical properties of composites," *Phys. Rev. B* **19**, 1783–1791 (1979).
74. G. Frye and R. L. Warnock, "Analysis of partial-wave dispersion relations," *Phys. Rev.* **130**, 478–494 (1963).
75. U. Fano and J. W. Cooper, "Spectral distribution of atomic oscillator strengths," *Rev. Mod. Phys.* **40**, 441–507 (1968).
76. P. Alippi, P. La Rocca, and G. B. Bachelet, "Alkali-metal plasmons, pseudopotentials, and optical sum rules," *Phys. Rev. B* **55**, 13835–13841 (1997).
77. F. Bassani and V. Lucarini, "General properties of optical harmonic generation from a simple oscillator model," *Nuovo Cimento* **20**, 1117–1125 (1998).
78. M. Altarelli and D. Y. Smith, "Superconvergence and sum rules for the optical constants: Physical meaning, comparison with experiment, and generalization," *Phys. Rev. B* **9**, 1290–1298 (1974).
79. J. T. Birmingham, S. M. Grannan, P. L. Richards, J. Kircher, M. Cardona, and A. Wittlin, "Optical absorptivity of $La_{1.87}Sr_{0.13}CuO_4$ below the superconducting plasma edge," *Phys. Rev. B* **59**, 647–653 (1999).
80. K.-E. Peiponen, P. Ketolainen, A. Vaittinen, and J. Riissanen, "Refractive index changes due to F→M conversion in KBr single crystals and their applications in holography," *Opt. Laser Technol.* **16**, 203–205 (1984).
81. B. W. Veal and A. P. Paulikas, "Optical properties of molybdenum. I. Experiment and Kramers-Kronig analysis," *Phys. Rev. B* **10**, 1280–1289 (1974).

82. C. W. Peterson and B. W. Knight, "Causality calculations in the time domain: An efficient alternative to the Kramers-Kronig method," *J. Opt. Soc. Am.* **63**, 1238–1242 (1973).
83. P. Lichvár, M. Liška, and D. Galusek, "What is true Kramers-Kronig transform," *Ceramics-Silikáty* **46**, 25–27 (2002).
84. K.-E. Peiponen, V. Lucarini, E. M. Vartiainen, and J. J. Saarinen, "Kramers-Kronig relations and sum rules of negative refractive index media," *Eur. Phys. J. B* **41**, 61–65 (2004).
85. V. G. Veselago, "Electrodynamics of substances with simultaneously negative electrical and magnetic susceptibilities," *Sov. Phys. Usp.* **10**, 509 (1968).
86. J. B. Pendry, "Negative refraction makes a perfect lens," *Phys. Rev. Lett.* **85**, 3966–3969 (2000).
87. Y. Zhang, B. Fluegel, and A. Mascarenhas, "Total negative refraction in real crystals for ballistic electrons and light," *Phys. Rev. Lett.* **91**, 157404 (2003).
88. B. Velicky, "Dispersion relation for complex reflectivity," *Chezh. J. Phys. B* **11**, 541–543 (1961).
89. H. Ehrenreich and H. R. Philipp, "Optical properties of Ag and Cu," *Phys. Rev.* **128**, 1622–1629 (1962).
90. H. Ehrenreich, "Electromagnetic transport in solids: optical properties and plasma effects," in *Proceedings of the International School of Physics Enrico Fermi*, J. Tauc, ed. (Academic Press, New York, 1966), pp. 106–154.
91. F. Wooten, *Optical Properties of Solids* (Academic Press, New York, 1972).
92. P. L. Nash, R. J. Bell, and R. Alexander, "On the Kramers-Kronig relation for the phase spectrum," *J. Mod. Opt.* **42**, 1837–1842 (1995).
93. M. H. Lee and O. I. Sindoni, "Kramers-Kronig relations with logarithmic kernel and application to the phase spectrum in the Drude model," *Phys. Rev. E* **56**, 3891–3896 (1997).
94. D. Y. Smith, "Dispersion relations and sum rules for magnetoreflectivity," *J. Opt. Soc. Am.* **66**, 547–554 (1976).
95. D. Y. Smith and C. A. Manogue, "Superconvergence relations and sum rules for reflection spectroscopy," *J. Opt. Soc. Am.* **71**, 935–947 (1981).
96. F. W. King, "Dispersion relations and sum rules for the normal reflectance of conductors and insulators," *J. Chem. Phys.* **71**, 4726–4733 (1979).
97. D. Y. Smith, "Dispersion relations for complex reflectivities," *J. Opt. Soc. Am.* **67**, 570–571 (1977).
98. J. S. Plaskett and P. N. Schatz, "On Robinson and Price (Kramers-Kronig) method of interpreting reflection data taken through a transparent window," *J. Chem. Phys.* **38**, 612 (1963).
99. S.-Y. Lee, "Phase recovery of the Raman amplitude from the Raman excitation profile," *Chem. Phys. Lett.* **245**, 620–628 (1995).
100. P. Grosse and V. Offermann, "Analysis of reflectance data using the Kramers-Kronig relations," *Appl. Phys. A* **52**, 138–144 (1991).
101. K. Yamamoto and H. Ishida, "Kramers-Kronig analysis of infrared reflection spectra with parallel polarization for isotropic materials," *Spectrochim. Acta* **50A**, 2079–2090 (1994).
102. K. Yamamoto and H. Ishida, "Complex refractive index determination for uniaxial anisotropy with the use of Kramers-Kronig analysis," *Appl. Spectrosc.* **51**, 1287–1293 (1997).
103. K. A. Fuller, H. D. Downing, and M. R. Querry, "Infrared optical properties of orthorhombic sulfur," *Appl. Opt.* **30**, 4081–4093 (1991).

104. F. A. Modine, R. W. Major, T. W. Haywood, G. R. Gruzalski, and D. Y. Smith, "Optical properties of tantalum carbide from the infrared to the near ultraviolet," *Phys. Rev. B* **29**, 836–841 (1984).
105. R. Hulthén, "Kramers-Kronig relations generalized - On dispersion-relations for finite frequency intervals - A spectrum-restoring filter," *J. Opt. Soc. Am.* **72**, 794–803 (1982).
106. G. W. Milton, D. J. Eyre, and J. V. Mantese, "Finite frequency range Kramers-Kronig relations: Bounds on the dispersion," *Phys. Rev. Lett.* **79**, 3062–3065 (1997).
107. K.-E. Peiponen and E. M. Vartiainen, "Kramers-Kronig relations in optical data inversion," *Phys. Rev. B* **44**, 8301–8303 (1991).
108. R. Z. Bachrach and F. C. Brown, "Exciton-optical properties of TlBr and TlCl," *Phys. Rev. B* **1**, 818–831 (1970).
109. R. K. Ahrenkiel, "Modified Kramers-Kronig analysis of optical spectra," *J. Opt. Soc. Am.* **61**, 1651–1655 (1971).
110. K. F. Palmer, M. Z. Williams, and B. A. Budde, "Multiply subtractive Kramers-Kronig analysis of optical data," *Appl. Opt.* **37**, 2660–2673 (1998).
111. V. Hughes and L. Grabner, "The radiofrequency spectrum of $Rb^{85}F$ and $Rb^{87}F$ by the electric resonance method," *Phys. Rev.* **79**, 314–322 (1950).
112. A. Battaglia, A. Gozzini, and E. Polacco, "On some phenomena related to the saturation of rotational resonances in the microwave spectrum of OCS," *Nuovo Cimento* **14**, 1076–1081 (1959).
113. T. W. Kaiser and C. G. B. Garrett, "Two-photon excitation in CaF_2:Eu^{2+}," *Phys. Rev. Lett.* **7**, 229–231 (1961).
114. P. A. Franken, A. E. Hill, C. W. Peters, and G. Weinrich, "Generation of optical harmonics," *Phys. Rev. Lett.* **7**, 118–119 (1961).
115. E. J. Woodbury and W. K. Ng, "Ruby laser operation in the near IR," *Proc. IRE.* **50**, 2367 (1962).
116. N. Bloembergen, "The stimulated Raman effect," *Am. J. Phys.* **35**, 989–1023 (1967).
117. G. L. Eesley, *Coherent Raman Spectroscopy* (Pergamon, Oxford, 1981).
118. P. D. Marker, R. W. Terhune, and C. M. Savage, "Optical third harmonic generation," in *Quantum Electronics*, P. Grivet and N. Bloembergen, eds. (Columbia University, New York, 1964).
119. S. Singh and L. Bradley, "Three-photon absorption in napthalene crystals by laser excitation," *Phys. Rev. Lett.* **12**, 612–614 (1964).
120. J. H. Hunt (ed.), *Selected Papers on Nonlinear Optical Spectroscopy* (SPIE Press, Bellingham, 2000).
121. F. J. Duarte (ed.), *Selected Papers on Dye Lasers* (SPIE Press, Bellingham, 1992).
122. V. Ter-Mikirtych (ed.), *Selected Papers on Tunable Solid-State Lasers* (SPIE Press, Bellingham, 2002).
123. A. L'Huillier, L. Lomprè, G. Mainfray, and C. Manus, "High order harmonic generation in gases," in *Atoms in Intense Laser Fields*, M. Gavrila, ed. (Academic, New York, 1991).
124. A. L'Huillier, M. Lewenstein, P. Saliéres, P. Balcou, M. Y. Ivanov, J. Larsson, and C.-G. Wahlström, "High-order harmonic-generation cutoff," *Phys. Rev. A* **48**, R3433–R3436 (1993).
125. C.-G. Wahlstrom, "High-order harmonic-generation using high-power lasers," *Phys. Scripta* **49**, 201–208 (1994).

126. R. Kubo, "Statistical-mechanical theory of irreversible processes. I. General theory and simple applications to magnetic and conduction problems," *J. Phys. Soc. Jpn.* **12**, 570–586 (1957).
127. F. Remacle and R. D. Levine, "Time domain information from resonant Raman excitation profiles: A direct inversion by maximum entropy," *J. Chem. Phys.* **99**, 4908–4925 (1993).
128. P. P. Kircheva and G. B. Hadjichristov, "Kramers-Kronig relations in FWM spectroscopy," *J. Phys. B* **27**, 3781–3793 (1994).
129. F. Bassani and V. Lucarini, "Pump and probe nonlinear processes: new modified sum rules from a simple oscillator model," *Eur. Phys. J. B.* **12**, 323–330 (1999).
130. D. A. Kleinman, "Laser and two-photon processes," *Phys. Rev.* **125**, 8788–8789 (1962).
131. C. J. Foot, B. Couillard, R. G. Beausoleil, and T. W. Hänsch, "On the electrodynamics of weakly nonlinear media," *Phys. Rev. Lett.* **54**, 1913–1916 (1985).
132. D. Fröhlich, "2-photon and 3-photon spectroscopy in solids," *Phys. Scripta* **T35**, 125–128 (1991).
133. A. J. Shields, M. Cardona, and K. Eberl, "Resonant Raman line shape of optic phonons in GaAs/AlAs multiple quantum wells," *Phys. Rev. Lett.* **72**, 412–415 (1994).
134. R. Wegerer, C. Thomsen, M. Cardona, H. J. Bornemann, and D. E. Morris, "Raman investigation of $YBa_{2-x}La_xCu_3O_7$ ceramics," *Phys. Rev. B* **53**, 3561–3565 (1996).
135. S. Jandl, T. Strach, T. Ruf, M. Cardona, V. Nevasil, M. Iliev, D. I. Zhigunov, S. N. Banio, and S. V. Shiryaev, "Raman study of crystal-field excitations in Pr_2CuO_4," *Phys. Rev. B* **56**, 5049–5052 (1997).
136. A. A. Sirenko, M. K. Zundekl, T. Ruf, K. Eberl, and M. Cardona, "Resonant Raman scattering in $InP/In_{0.48}Ga_{0.52}P$ quantum dot structures embedded in a waveguide," *Phys. Rev. B* **58**, 12633–12636 (1998).
137. D. Fröhlich, A. Nöthe, and K. Reimann, "Observation of the resonant optical Stark effect in a semiconductor," *Phys. Rev. Lett.* **55**, 1335–1337 (1985).
138. D. Fröhlich, R. Wille, W. Schlapp, and G. Weimann, "Optical quantum-confined Stark effect in GaAs quantum wells," *Phys. Rev. Lett.* **59**, 1748–1751 (1987).
139. W. H. Knox, D. S. C. D. A. B. Miller, and S. Schmitt-Rink, "Femtosecond AC Stark-effect in semiconductor quantum wells - extreme low-intensity and high-intensity limits," *Phys. Rev. Lett.* **62**, 1189–1192 (1989).
140. D. Fröhlich, C. Neumann, B. Uebbing, and R. Wille, "Experimental investigation of 3-level optical Stark-effect in semiconductors," *Phys. Stat. Sol. (b)* **159**, 297–307 (1990).
141. D. Fröhlich, S. Spitzer, B. Uebbing, and R. Zimmermann, "Polarization dependence in the intraband optical Stark-effect of quantum-wells," *Phys. Stat. Sol. (b)* **173**, 83–89 (1992).
142. H. Kishida, H. Matsuzaki, H. Okamoto, T. Manabe, M. Yamashita, Y. Taguchi, and Y. Tokura, "Gigantic optical nonlinearity in one-dimensional Mott-Hubbard insulators," *Nature* **405**, 929–932 (2000).

143. H. Kishida, M. Ono, K. Miura, H. Okamoto, M. Izumi, T. Manako, M. Kawasaki, Y. Taguchi, Y. Tokura, T. Tohyama, K. Tsutsui, and S. Maekawa, "Large third-order optical nonlinearity of Cu-O chains investigated by third harmonic generation spectroscopy," *Phys. Rev. Lett.* **87**, 177401 (2001).
144. J. E. Sipe and R. W. Boyd, "Nonlinear susceptibility of composite optical materials in the Maxwell Garnett model," *Phys. Rev. A* **46**, 1614–1629 (1992).
145. J. W. Haus, H. S. Zhou, S. Takami, M. Hirasawa, I. Honma, and H. Komiyama, "Enhanced optical properties of metal-coated nanoparticles," *J. Appl. Phys.* **73**, 1043–1048 (1993).
146. G. L. Fischer, R. W. Boyd, R. J. Gehr, S. A. Jenekhe, J. A. Osaheni, J. E. Sipe, and L. A. Weller-Brophy, "Enhanced nonlinear optical response of composite materials," *Phys. Rev. Lett.* **74**, 1871–1874 (1995).
147. D. Faccio, P. Di Trapani, E. Borsella, F. Gonella, P. Mazzoldi, and A. M. Malvezzi, "Measurement of the third-order nonlinear susceptibility of Ag nanoparticles in glass in a wide spectral range," *Europhys. Lett.* **43**, 213–218 (1998).
148. V. M. Shalaev, E. Y. Poliakov, and V. A. Markel, "Small-particle composites. II. Nonlinear optical properties," *Phys. Rev. B* **53**, 2437–2449 (1996).
149. D. J. Bergman, O. Levy, and D. Stroud, "Theory of optical bistability in a weakly nonlinear composite medium," *Phys. Rev. B* **49**, 129–134 (1994).
150. Y.-K. Yoon, R. S. Bennink, R. W. Boyd, and J. E. Sipe, "Intrinsic optical bistability in a thin layer of nonlinear optical material by means of local field effects," *Opt. Commun.* **179**, 577–580 (2000).
151. J. W. Haus, R. Inguva, and C. M. Bowden, "Effective-medium theory of nonlinear ellipsoidal composites," *Phys. Rev. A* **40**, 5729–5734 (1989).
152. J. W. Haus, N. Kalyaniwalla, R. Inguva, M. Bloemer, and C. M. Bowden, "Nonlinear-optical properties of conductive spheroidal particle composites," *J. Opt. Soc. Am. B* **6**, 797–807 (1989).
153. D. D. Smith, G. Fischer, R. W. Boyd, and D. A. Gregory, "Cancellation of photoinduced absorption in metal nanoparticle composites through a counterintuitive consequence of local field effects," *J. Opt. Soc. Am. B* **14**, 1625–1631 (1997).
154. A. E. Neeves and M. H. Birnboim, "Composite structures for the enhancement of nonlinear-optical susceptibility," *J. Opt. Soc. Am. B* **6**, 787–796 (1989).
155. P. Hänninen, A. Soini, N. Meltola, J. Soini, J. Soukka, and E. Soini, "A new microvolume technique for bioaffinity assays using two-photon excitation," *Nat. Biotechnology* **18**, 548–550 (2000).
156. A. Lakhtakia, "Application of strong permittivity fluctuation theory for isotropic, cubically nonlinear, composite mediums," *Opt. Commun.* **192**, 145–151 (2001).
157. B. Yang, C. Zhang, Q. Wang, Z. Zhang, and D. Tian, "Enhancement of optical nonlinearity for periodic anisotropic composites," *Opt. Commun.* **183**, 307–315 (2000).
158. L. Gao, K. W. Yu, Z. Y. Li, and B. Hu, "Effective nonlinear optical properties of metal-dielectric composite media with shape distribution," *Phys. Rev. E* **64**, 036615 (2001).
159. R. J. Gehr, G. L. Fischer, R. W. Boyd, and J. E. Sipe, "Nonlinear optical response of layered composite materials," *Phys. Rev. A* **53**, 2792–2798 (1996).

160. R. L. Nelson and R. W. Boyd, "Enhanced electro-optic response of layered composite materials," *Appl. Phys. Lett.* **74**, 2417–2419 (1999).
161. S. Abe, M. Schneider, and W.-P. Su, "Nonlinear optical susceptibilities of polysilanes: exciton effect," *Chem. Phys. Lett.* **192**, 425–429 (1992).
162. S. Abe, M. Schreiber, W. P. Su, and J. Yu, "Excitons and nonlinear optical spectra in conjugated polymers," *Phys. Rev. B* **45**, 9432–9435 (1992).
163. J. J. Saarinen, E. M. Vartiainen, and K.-E. Peiponen, "On tailoring of nonlinear spectral properties of nanocomposites having Maxwell Garnett or Bruggeman structure," *Opt. Rev.* **10**, 111–115 (2003).
164. J. Joseph and A. Gagnaire, "Ellipsometric study of anodic oxide growth: Application to the titanium oxide systems," *Thin Solid Films* **103**, 257–265 (1983).
165. T. Hasegawa, Y. Iwasa, T. Koda, H. Kishida, Y. Tokura, S. Wada, H. Tashiro, H. Tachibana, and M. Matsumoto, "Nature of one-dimensional excitons in polysilanes," *Phys. Rev. B* **54**, 11365–11374 (1996).
166. W. E. Torruellas, D. Neher, R. Zanoni, G. I. Stegeman, F. Kajzar, and M. Leclerc, "Dispersion measurements of the third-order nonlinear susceptibility of polythiophene thin films," *Chem. Phys. Lett.* **175**, 11–16 (1990).
167. S. Mittler-Neher, "Linear optical characterization of nonlinear optical polymers for integrated optics," *Macromol. Chem. Phys.* **199**, 513–523 (1998).
168. W. Loong and H. Pan, "A direct approach to the modeling of polydihexylsilane as a contrast enhancement material," *J. Vac. Sci. Technol.* **8**, 1731–1734 (1990).
169. B. J. Orr and J. F. Ward, "Perturbation theory of the non-linear optical polarization of an isolated system," *Mol. Phys.* **20**, 513–526 (1971).
170. P. J. Price, "Theory of quadratic response functions," *Phys. Rev.* **130**, 1792–1797 (1963).
171. M. Kogan, "On the electrodynamics of weakly nonlinear media," *Sov. Phys. JETP* **16**, 217–219 (1963).
172. W. J. Caspers, "Dispersion relations for nonlinear response," *Phys. Rev.* **133**, 1249–1251 (1964).
173. F. L. Ridener and R. H. Good, Jr., "Dispersion relations for third-degree nonlinear systems," *Phys. Rev. B* **10**, 4980–4987 (1974).
174. F. L. Ridener and R. H. Good, Jr., "Dispersion relations for nonlinear systems of arbitrary degree," *Phys. Rev. B* **11**, 2768–2770 (1975).
175. E. Ghahramani, D. Moss, and J. E. Sipe, "Full-band-structure calculation of first-, second-, and third-harmonic optical response coefficients of ZnSe, ZnTe, and CdTe," *Phys. Rev. B* **43**, 9700–9710 (1991).
176. L. Kador, "Kramers-Kronig relations in nonlinear optics," *Appl. Phys. Lett.* **66**, 2938–2939 (1995).
177. J. L. P. Hughes, Y. Wang, and J. E. Sipe, "Calculation of linear and second-order optical response in wurtzite GaN and AlN," *Phys. Rev. B* **55**, 13630–13640 (1997).
178. J. E. Sipe and E. Ghahramani, "Nonlinear optical response of semiconductors in the independent-particle approximation," *Phys. Rev. B* **48**, 11705–11722 (1993).
179. M. Sheik-Bahae, "Nonlinear optics of bound electrons in solids," in *Nonlinear Optical Materials*, J. V. Moloney, ed. (Springer, New York, 1998).
180. S. Scandolo and F. Bassani, "Kramers-Kronig relations and sum rules for the second-harmonic susceptibility," *Phys. Rev. B* **51**, 6925–6927 (1995).

181. V. Chernyak and S. Mukamel, "Generalized sum rules for optical nonlinearities of many electron system," *J. Chem. Phys.* **103**, 7640–7644 (1995).
182. J. P. Burg, *Maximum Entropy Spectral Analysis*, Ph.D. thesis (Stanford University, Department of Geophysics, 1975).
183. E. M. Vartiainen, T. Asakura, and K.-E. Peiponen, "Generalized noniterative maximum entropy procedure for phase retrieval problems in optical spectroscopy," *Opt. Commun.* **104**, 149–156 (1993).
184. E. M. Vartiainen, K.-E. Peiponen, and T. Asakura, "Sum rules in testing nonlinear susceptibility obtained using the maximum entropy model," *J. Phys.: Condens. Matter* **5**, L113–L116 (1993).
185. E. M. Vartiainen and K.-E. Peiponen, "Meromorphic degenerate nonlinear susceptibility: Phase retrieval from the amplitude spectrum," *Phys. Rev. B* **50**, 1941–1944 (1994).
186. E. M. Vartiainen, K.-E. Peiponen, and T. Asakura, "Dispersion theory and phase retrieval of meromorphic total susceptibility," *J. Phys.: Condens. Mat.* **9**, 8937–8943 (1997).
187. S. E. Harris, "Electromagnetically induced transparency," *Phys. Today* **50**, 36–42 (1997).
188. J. J. Saarinen, "Sum rules for arbitrary-order harmonic generation susceptibilities," *Eur. Phys. J. B* **30**, 551–557 (2002).
189. F. Bassani and V. Lucarini, "Asymptotic behaviour and general properties of harmonic generation susceptibilities," *Eur. Phys. J. B* **17**, 567–573 (2000).
190. N. P. Rapapa and S. Scandolo, "Universal constraints for the third-harmonic generation susceptibility," *J. Phys.: Condens. Matter* **8**, 6997–7004 (1996).
191. F. W. King, "Efficient numerical approach to the evaluation of Kramers-Kronig transforms," *J. Opt. Soc. Am. B* **19**, 2427–2436 (2002).
192. C. Y. Fong and Y. R. Shen, "Theoretical studies on the dispersion of the nonlinear optical susceptibilities in GaAs, InAs, and InSb," *Phys. Rev. B* **12**, 2325–2328 (1975).
193. D. J. Moss, J. E. Sipe, and H. M. Van Driel, "Empirical tight-binding calculation of dispersion in the second-order nonlinear optical constant for zinc-blende crystals," *Phys. Rev. B* **36**, 9708–9721 (1987).
194. C. G. B. Garrett and F. N. G. Robinson, "Miller's phenomenological rule for computing nonlinear susceptibilities," *IEEE J. Quantum Electron.* **2**, 328–329 (1966).
195. E. Tokunaga, A. Terasaki, and T. Kobayashi, "Femtosecond time-resolved dispersion relations studied with a frequency-domain interferometer," *Phys. Rev. A* **47**, 4581–4584 (1993).
196. K.-E. Peiponen, J. J. Saarinen, and Y. Svirko, "Derivation of general dispersion relations and sum rules for meromorphic nonlinear optical spectroscopy," *Phys. Rev. A* **69**, 043818 (2004).
197. R. Nevanlinna and V. Paatero, *Introduction to Complex Analysis* (Chelsea, New York, 1964).
198. F. W. King, "Sum rules for the optical constants," *J. Math. Phys.* **17**, 1509–1514 (1976).
199. E. M. Vartiainen, K.-E. Peiponen, and T. Asakura, "Maximum entropy model in reflection spectra analysis," *Opt. Commun.* **89**, 37–40 (1992).
200. J. K. Kauppinen, D. J. Moffatt, M. R. Hollberg, and H. H. Mantsch, "A new line-narrowing procedure based on Fourier self-deconvolution, maximum entropy, and linear prediction,," *Appl. Spectrosc.* **45**, 411 (1991).

201. J. K. Kauppinen, D. J. Moffatt, M. R. Hollberg, and H. H. Mantsch, "Characteristics of the LOMEP line-narrowing method," *Appl. Spectrosc.* **45**, 1516 (1991).
202. E. M. Vartiainen, K.-E. Peiponen, and T. Asakura, "Comparison between the optical constants obtained by the Kramers-Kronig analysis and the maximum entropy method: infrared optical properties of orthorhompic sulfur," *Appl. Opt.* **32**, 1126–1129 (1993).
203. E. M. Vartiainen, K.-E. Peiponen, and T. Asakura, "Phase retrieval in optical spectroscopy: Resolving optical constants from power spectra," *Appl. Spectrosc.* **50**, 1283–1289 (1996).
204. J. A. Räty, E. M. Vartiainen, and K.-E. Peiponen, "Resolving optical constants from reflectance of liquids in the UV-visible range," *Appl. Spectrosc.* **53**, 92–96 (1999).
205. E. M. Vartiainen, "Phase retrieval approach for coherent anti-Stokes Raman scattering spectrum analysis," *J. Opt. Soc. Am. B* **9**, 1209–1214 (1992).
206. E. M. Vartiainen, K.-E. Peiponen, H. Kishida, and T. Koda, "Phase retrieval in nonlinear optical spectroscopy by the maximum-entropy method: an application to the $|\chi^{(3)}|$ spectra of polysilane," *J. Opt. Soc. Am. B* **13**, 2106–2114 (1996).
207. P.-K. Yang and J. Y. Huang, "Phase-retrieval problems in infrared-visible sum-frequency generation spectroscopy by the maximum-entropy model," *J. Opt. Soc. Am. B* **14**, 2443–2448 (1997).
208. P.-K. Yang and J. Y. Huang, "Model-independent maximum-entropy method for the analysis of sum-frequency vibrational spectroscopy," *J. Opt. Soc. Am B* **17**, 1216–1222 (2000).
209. K.-E. Peiponen, E. M. Vartiainen, and T. Asakura, "Dispersion theory and phase retrieval of meromorphic total susceptibility," *J. Phys. C.: Condens. Matter* **9**, 8937–8943 (1997).
210. K.-E. Peiponen, E. M. Vartiainen, and T. Asakura, "Dispersion theory of effective meromorphic nonlinear susceptibilities of nanocomposites," *J. Phys.: Condens. Matter* **10**, 2483–2488 (1998).
211. S. Haykin and S. Kesler, "Prediction-error filtering and maximum entropy spectral estimation," in *Nonlinear Methods of Spectral Analysis*, 2nd ed., S. Haykin, ed. (Springer, Berlin, 1983), pp. 9–72.
212. J.-F. Brun, D. D. S. Meneses, B. Rousseau, and P. Echegut, "Dispersion relations and phase retrieval in infrared reflection spectra analysis," *Appl. Spectrosc.* **55**, 774–780 (2001).
213. J. Räty and K.-E. Peiponen, "Measurement of refractive index of liquids using s- and p-polarized light," *Meas. Sci. Technol.* **11**, 74–76 (2000).
214. J. Räty, K.-E. Peiponen, and E. Keränen, "The complex refractive index measurement of liquids by a novel reflectometer apparatus for the UV–visible spectral range," *Meas. Sci. Technol.* **9**, 95–99 (1998).
215. K.-E. Peiponen, A. J. Jääskeläinen, E. M. Vartiainen, J. Räty, U. Tapper, O. Richard, E. I. Kauppinen, and K. Lumme, "Estimation of wavelength-dependent effective refractive index of spherical plastic pigments in a liquid matrix," *Appl. Opt.* **40**, 5482–5486 (2001).
216. E. M. Vartiainen, K.-E. Peiponen, and T. Asakura, "Retrieval of optical constants from magnetoreflectance by maximum entropy model," *Opt. Rev.* **5**, 271–274 (1998).

217. J. Räty, K.-E. Peiponen, and T. Asakura, *UV-Visible Reflection Spectroscopy of Liquids* (Springer, Heidelberg, 2004).
218. H. Räther, *Surface Plasmons on Smooth and Rough Surfaces and on Gratings* (Springer, Berlin, 1988).
219. J. Homola, S. S. Yee, and G. Gauglitz, "Surface plasmon resonance sensors: review," *Sensor Actuat. B* **54**, 3–15 (1999).
220. G. J. Sprokel and J. D. Swalen, "Attenuated total reflection method," in *Handbook of Optical Constants of Solids*, E. D. Palik, ed. (Academic Press, New York, 1998).
221. R. M. A. Azzam and N. M. Bashara, *Ellipsometry and Polarized Light* (North-Holland, Amsterdam, 1977).
222. W. P. Chen and J. M. Chen, "Surface plasma wave study of submonolayer Cs and CsO covered Ag surfaces," *Surf. Sci.* **91**, 601 (1980).
223. W. P. Chen and J. M. Chen, "Use of surface plasma waves for determination of the thickness and optical constants of thin metallic films," *J. Opt. Soc. Am.* **71**, 189–191 (1981).
224. B. G. Tilkens, Y. F. Lion, and Y. L. Renotte, "Uncertainties in the values obtained by surface plasmon resonance," *Opt. Eng.* **39**, 363 (2000).
225. C. Nylander, B. Liendberg, and T. Lind, "Gas detection by means of surface plasmon resonance," *Sensor Actuat.* **3**, 79–88 (1982/83).
226. L. A. Vanderberg, "Detection of biological agents: Looking for bugs in all the wrong places," *Appl. Spectrosc.* **54**, 376A–385A (2000).
227. R. J. Green, R. A. Frazier, K. M. Skakesheff, M. C. Davies, C. J. Roberts, and S. J. B. Tendler, "Surface plasmon resonance analysis of dynamic biological interactions with biomaterials," *Biomaterials* **21**, 1823–1835 (2000).
228. S. Löfås, M. Malmqvist, I. Rönnberg, E. Stenberg, B. Liedberg, and I. Lundström, "Bioanalysis with surface plasmon resonance," *Sensor Actuat. B* **5**, 79–84 (1991).
229. K. Matsubara, S. Kawata, and S. Minami, "Optical chemical sensor based on surface plasmon measurement," *Appl. Opt.* **27**, 1160–1163 (1988).
230. L.-M. Zhang and D. Uttamchandani, "Optical chemical sensing employing surface plasmon resonance," *Electron. Lett.* **24**, 1469–1470 (1988).
231. J. J. Saarinen, K.-E. Peiponen, and E. M. Vartiainen, "Simulation on wavelength-dependent complex refractive index of liquids obtained by phase retrieval from reflectance dip due to surface plasmon resonance," *Appl. Spectrosc.* **57**, 288–292 (2003).
232. J. J. Saarinen, E. M. Vartiainen, and K.-E. Peiponen, "Retrieval of the complex permittivity of spherical nanoparticles in a liquid host material from a spectral surface plasmon resonance measurement," *Appl. Phys. Lett.* **83**, 893–895 (2003).
233. M. C. Nuss and J. Orenstein, "Terahertz time-domain spectroscopy," in *Millimeter and Submillimeter Wave Spectroscopy of Solids*, G. Grüner, K. Dahl, and C. Dahl, eds. (Springer, Berlin, 1998), pp. 7–50.
234. S. C. Howells and L. A. Schlie, "Transient terahertz reflection spectroscopy of undoped InSb from 0.1 to 1.1 THz," *Appl. Phys. Lett.* **69**, 550–552 (1996).
235. T.-I. Jeon and D. Grischkowsky, "Characterization of optically dense, doped semiconductors by reflection THz time domain spectroscopy," *Appl. Phys. Lett.* **72**, 3032–3034 (1998).

236. R. Shimano, Y. Ino, Y. P. Svirko, and M. Kuwata-Gonokami, "Terahertz frequency Hall measurement by magneto-optical Kerr spectroscopy in InAs," *Appl. Phys. Lett.* **81**, 199–201 (2002).
237. S. Nashima, O. Morikawa, K. Takata, and M. Hangyo, "Measurement of optical properties of highly doped silicon by terahertz time domain reflection spectroscopy," *Appl. Phys. Lett.* **79**, 3923–3925 (2001).
238. M. Kazan, R. Meissner, and I. Wilke, "Convertible transmission-reflection time-domain terahertz spectrometer," *Rev. Sci. Instrum.* **72**, 3427–3430 (2001).
239. E. M. Vartiainen, Y. Ino, R. Shimano, M. Kuwata-Gonokami, Y. P. Svirko, and K.-E. Peiponen, "Numerical phase correction method for terahertz time-domain reflection spectroscopy," *J. Appl. Phys.* **96**, 4171–4175 (2004).

Index

absorption 2, 14, 17, 21, 27–29, 32, 35, 47, 76, 79, 124
anchor point 44, 47, 90, 92, 93, 95, 102–104, 115–117, 119, 120, 124, 128, 129, 131
angle of incidence 14, 117, 119, 123–125
ATR 121, 122
Attenuated Total Reflection *see* ATR

Beer-Lambert law 1, 27
birefringence 25
Blaschke product 41
Brewster angle 41
Bruggeman 21–24, 30, 115, 129
  effective medium 23, 24
  liquid 115, 129

Cauchy 76
causality 1, 27, 28, 33, 72, 74, 75, 89, 109, 133
Clausius-Mossotti equation 20, 21
conductors 29, 30, 33, 35, 38, 39, 89, 90, 133
convergence 8, 45, 47, 48, 81, 95–100, 103

density matrix 9, 18, 79, 87
dielectric function 2, 12, 16, 22, 23, 26, 32, 33, 36, 37, 41, 120, 122–124, 127, 128, 131
  effective 21, 23–25, 30, 125, 126
dispersion relations 3, 27–30, 35, 39, 40, 71, 73, 75, 77, 78, 87, 88, 96, 97, 109, 110, 112, 133, 134

ellipsometry 42, 44, 117
error 44, 47, 48, 90, 92, 95, 96, 102, 104, 106, 116, 117, 119, 126–128
  experimental 96
  phase 126–128
  truncation 104, 106
extinction coefficient 35, 43, 44, 117, 118, 120, 121
extrapolation 44, 93

field
  electromagnetic 5–7, 10, 19
  local 11, 19, 20, 30, 31, 133
Fresnel's equations 15, 39, 117

Green function 11, 16–18, 29, 30, 72, 74, 85
  linear 11, 16, 17, 29
  nonlinear 72, 74, 85

Hamiltonian 6–9, 16, 17, 30, 85, 87
harmonic-generation 2, 79, 83–90, 93–98, 101, 102, 106, 133, 134
  processes 83, 84, 90, 133
  susceptibility 79, 83–90, 93–98, 101, 102, 106
holomorphic function 109

K-K relation 2
K-K relations 2, 3, 27, 29–31, 33–36, 39, 42, 44–48, 71–73, 75–80, 83, 84, 87–90, 92–98, 101–104, 109, 111, 128, 129, 134, 137
Kramers-Kronig Relations *see* K-K relations
Kramers-Kronig relations
  Multiply Subtractive *see* MSKK
  Singly Subtractive *see* SSKK

Lagrangian 6, 7
laser 2, 3, 73, 76–79, 126, 129

linear susceptibility 12, 16–18, 20, 25, 29–35, 46, 48, 78, 86, 106, 107
  effective 25

MATLAB 3, 137
Maximum Entropy Method *see* MEM
Maxwell Garnett system *see* MG
MEM 115–117, 119–121, 124–126, 128–131, 134
MG 21–26, 30, 34, 124, 125, 128
MSKK 47, 48, 72, 83, 90, 91, 104

nanocomposite 23, 128
nanostructure 3, 21–27, 30, 31, 33, 46, 135
  layered 22, 24–26, 30, 46
nonlinear susceptibility 71–75, 77–79, 92, 96, 101, 102, 104, 106, 109, 110, 115, 117, 128–130, 133, 134
  degenerate 115, 129
  effective 129, 130
  holomorphic 109, 128
  meromorphic 109, 134

optics 2, 19, 23, 27, 33, 71–73, 79, 83, 89, 90, 92, 109, 110, 112, 114, 133, 134
  linear 2, 19, 23, 27, 33, 79, 89, 90, 109, 114, 133
  nonlinear 71–73, 83, 90, 92, 109, 110, 112, 134
oscillator
  anharmonic 87, 105, 107
  Drude-Lorentz 33, 120

paper 119, 120
phase retrieval 2, 3, 39–42, 47, 115–117, 124, 128, 134, 135
polarization 5, 6, 9–13, 16, 19, 20, 29, 30, 41, 77, 84, 85, 117, 122
  linear 11, 12, 16
  macroscopic 19, 20
  microscopic 19
  nonlinear 77, 84, 85, 117
pole 28–30, 76, 77, 89, 90, 109–111, 114, 134
polysilane 94, 96, 97, 99, 101, 102, 115, 128
polythiophene 93, 95–98, 100, 101
prism 24, 117, 119, 122–124
pump-and-probe 72–77, 79–81, 89, 104, 109, 133
  susceptibility 72, 74–77, 79, 80

reflectance 2, 15, 23, 39–44, 47, 48, 115–119, 121–124, 127–129, 131, 133, 134
reflectivity 2, 24, 34, 35, 39–42, 46, 48, 109, 115–117, 128, 131
refractive index 1, 25, 28, 35, 36, 39, 41–44, 47, 48, 117–121, 123–126
  negative 39
residue 110–114
resonance 23, 79, 92, 103, 106, 111, 120, 122–124, 128

spectroscopy
  reflectance 39
  reflection 2, 42, 47, 115, 117
  terahertz 3, 126
  transmission 1, 2
SPR 122–125
squeezing procedure 120
SSKK 47, 90, 92, 101–104, 137
sum rule 101
sum rules 1–3, 27, 31, 33–35, 39, 45, 46, 71, 72, 78–80, 83, 84, 87–90, 93, 98–101, 106, 109, 110, 112–114, 117, 133, 134
surface plasmon resonance *see* SPR

TE-polarized light 14, 24, 25, 41, 119
tensor 11, 12, 18, 30, 74, 77, 83, 89
theorem
  Scandolo 73–76, 83, 84, 110, 133
  superconvergence 31–34, 36, 45, 78, 80, 88, 90, 133, 134
  Titchmarsch 28, 29, 72, 74, 133
TM-polarized light 14, 15, 25, 26, 41, 42, 117, 122
Toeplitz system 116
transform
  Fourier 5, 11, 12, 16–18, 28, 39, 85, 116, 126
  Hilbert 2, 28–30, 74, 110, 111

Springer Series in

OPTICAL SCIENCES

Volume 1

1 **Solid-State Laser Engineering**
By W. Koechner, 5th revised and updated ed. 1999, 472 figs., 55 tabs., XII, 746 pages

Published titles since volume 80

80 **Optical Properties of Photonic Crystals**
By K. Sakoda, 2nd ed., 2004, 107 figs., 29 tabs., XIV, 255 pages

81 **Photonic Analog-to-Digital Conversion**
By B.L. Shoop, 2001, 259 figs., 11 tabs., XIV, 330 pages

82 **Spatial Solitons**
By S. Trillo, W.E. Torruellas (Eds), 2001, 194 figs., 7 tabs., XX, 454 pages

83 **Nonimaging Fresnel Lenses**
Design and Performance of Solar Concentrators
By R. Leutz, A. Suzuki, 2001, 139 figs., 44 tabs., XII, 272 pages

84 **Nano-Optics**
By S. Kawata, M. Ohtsu, M. Irie (Eds.), 2002, 258 figs., 2 tabs., XVI, 321 pages

85 **Sensing with Terahertz Radiation**
By D. Mittleman (Ed.), 2003, 207 figs., 14 tabs., XVI, 337 pages

86 **Progress in Nano-Electro-Optics I**
Basics and Theory of Near-Field Optics
By M. Ohtsu (Ed.), 2003, 118 figs., XIV, 161 pages

87 **Optical Imaging and Microscopy**
Techniques and Advanced Systems
By P. Török, F.-J. Kao (Eds.), 2003, 260 figs., XVII, 395 pages

88 **Optical Interference Coatings**
By N. Kaiser, H.K. Pulker (Eds.), 2003, 203 figs., 50 tabs., XVI, 504 pages

89 **Progress in Nano-Electro-Optics II**
Novel Devices and Atom Manipulation
By M. Ohtsu (Ed.), 2003, 115 figs., XIII, 188 pages

90/1 **Raman Amplifiers for Telecommunications 1**
Physical Principles
By M.N. Islam (Ed.), 2004, 488 figs., XXVIII, 328 pages

90/2 **Raman Amplifiers for Telecommunications 2**
Sub-Systems and Systems
By M.N. Islam (Ed.), 2004, 278 figs., XXVIII, 420 pages

91 **Optical Super Resolution**
By Z. Zalevsky, D. Mendlovic, 2004, 164 figs., XVIII, 232 pages

92 **UV-Visible Reflection Spectroscopy of Liquids**
By J.A. Räty, K.-E. Peiponen, T. Asakura, 2004, 131 figs., XII, 219 pages

93 **Fundamentals of Semiconductor Lasers**
By T. Numai, 2004, 166 figs., XII, 264 pages

94 **Photonic Crystals**
Physics, Fabrication and Applications
By K. Inoue, K. Ohtaka (Eds.), 2004, 209 figs., XV, 320 pages

95 **Ultrafast Optics IV**
Selected Contributions to the 4th International Conference
on Ultrafast Optics, Vienna, Austria
By F. Krausz, G. Korn, P. Corkum, I.A. Walmsley (Eds.), 2004, 281 figs., XIV, 506 pages

Springer Series in
OPTICAL SCIENCES

96 **Progress in Nano-Electro Optics III**
Industrial Applications and Dynamics of the Nano-Optical System
By M. Ohtsu (Ed.), 2004, 186 figs., 8 tabs., XIV, 224 pages

97 **Microoptics**
From Technology to Applications
By J. Jahns, K.-H. Brenner, 2004, 303 figs., XI, 335 pages

98 **X-Ray Optics**
High-Energy-Resolution Applications
By Y. Shvyd'ko, 2004, 181 figs., XIV, 404 pages

99 **Mono-Cycle Photonics and Optical Scanning Tunneling Microscopy**
Route to Femtosecond Ångstrom Technology
By M. Yamashita, H. Shigekawa, R. Morita (Eds.) 2005, 241 figs., XX, 393 pages

100 **Quantum Interference and Coherence**
Theory and Experiments
By Z. Ficek and S. Swain, 2005, 178 figs., approx. 432 pages

101 **Polarization Optics in Telecommunications**
By J. Damask, 2005, 110 figs., XVI, 528 pages

102 **Lidar**
Range-Resolved Optical Remote Sensing of the Atmosphere
By C. Weitkamp (Ed.), 161 figs., approx. 416 pages

103 **Optical Fiber Fusion Splicing**
By A.D. Yablon, 2005, 100 figs., approx. IX, 310 pages

104 **Optoelectronics of Molecules and Polymers**
By A. Moliton, 2005, 200 figs., approx. 460 pages

105 **Solid-State Random Lasers**
By M. Noginov, 2005, 149 figs., approx. XII, 380 pages

106 **Coherent Sources of XUV Radiation**
Soft X-Ray Lasers and High-Order Harmonic Generation
By P. Jaeglé, 2005, 150 figs., approx. 264 pages

107 **Optical Frequency-Modulated Continuous-Wave (FMCW) Interferometry**
By J. Zheng, 2005, 137 figs., approx. 250 pages

108 **Laser Resonators and Beam Propagation**
Fundamentals, Advanced Concepts and Applications
By N. Hodgson and H. Weber, 2005, 497 figs., approx. 790 pages

109 **Progress in Nano-Electro Optics IV**
Characterization of Nano-Optical Materials and Optical Near-Field Interactions
By M. Ohtsu (Ed.), 2005, 123 figs., XIV, 206 pages

110 **Kramers–Kronig Relations in Optical Materials Research**
By V. Lucarini, J.J. Saarinen, K.-E. Peiponen, E.M. Vartiainen, 2005, 37 figs., approx. IX, 170 pages

111 **Semiconductor Lasers**
Stability, Instability and Chaos
By J. Ohtsubo, 2005, 169 figs., approx. XII, 438 pages

112 **Photovoltaic Solar Energy Generation**
By A. Goetzberger and V.U. Hoffmann, 2005, 138 figs., approx. IX, 245 pages

Printing: Mercedes-Druck, Berlin
Binding: Stein+Lehmann, Berlin